"In her compelling memoir . . . Muir discovers that Hemmed's work and the inventiveness of her father in particular was grounded in something other than weapon-making."—*Jerusalem Post*

"*The Book of Telling* tells of a woman's journey to uncover the secret life of her father and to find herself in the process, an unusual counterpoint between personal history and the history of a young nation. Haunting, powerful, and beautifully written."— Alan Lightman, author of *Einstein's Dreams*

"Sharona Muir has written a gripping personal memoir about her odyssey to rediscover and reclaim her father. Along the way she uncovers some hard truths about the heroic founders of Israel and the beginnings of Israeli science. *The Book of Telling* keeps in all the fears and resentments and consolations and warmth of such a process—at once her own story and the tale of a nation."—Edmund White, author of *A Boy's Own Story*

"In the rich field of contemporary Jewish American female memoir, this book is among the finest in print. With this daughter's journey come remarkable insights into the nature of genius, the nature of cultural dislocation, the inventions of new identity, memory, femininity, and history. Identity, female identity included, as Muir painstakingly unveils it in this novel of origins, is inseparable from world history, personal memory, postmodern diasporas, the age of science, and the ever-present threat of Jewish extinction."—Gloria Cronin, editor of *Saul Bellow Journal*

"An extraordinary story, exceptionally well told, and absolutely true to character. I met people like these during my years in Israel, painful amalgams of irrepressible brilliance and unconquerable melancholy who would sometimes allude to a mysterious past but seldom elaborated. Sharona Muir has done so well in getting them to talk while, at the same time, bringing out their faults and human flaws."— Norman Lebrecht, author of *The Song of Names*

"*The Book of Telling* opens with a sequence of lyric evocations of the elusive father whose influence made Sharona Muir into both a poet and a scholar. By the end of this memoir, her passionate investigation has drawn Itzhak Bentov partly out of the shadows that protected his work as an Israeli defense scientist and given the book a historical scope that never ceases to be poignantly intimate."—Diane Middlebrook, author of *Her Husband: Sylvia Plath and Ted Hughes—A Marriage*

"A fascinating narrative, both poetic and sober, appreciative of complexities and free of self-delusions, working from sign to symbol, through personal experience and second-hand testimony to flashes of imaginative reconstruction and existential insight."—Leona Toker, Hebrew University, Jerusalem

The Book of Telling

Tracing the Secrets of My Father's Lives

SHARONA BEN-TOV MUIR

With a new preface by the author

University of Nebraska Press
Lincoln and London

Library of Congress Cataloging-in-Publication Data
Muir, Sharona, 1957–
The book of telling: tracing the secrets of my father's lives /
Sharona Ben-Tov Muir; with a new preface by the author.
p. cm.
Includes bibliographical references.
ISBN 978-0-8032-1648-8 (pbk.: alk. paper)
1. Bentov, Itzhak.   2. Rocketry—Israel—Biography.   3. Philosophers—
Israel—Biography.   4. Muir, Sharona, 1957—Childhood and youth.
5. Fathers and daughters—Israel.   I. Title.
TL781.85.B46M85 2008
621.43'56092—dc22
[B]
2007034922

Preface

Despite appearances, the heart of this book is not Israel's secrets, family secrets, fathers and daughters, Itzhak Bentov, missiles, pearls, plane crashes, gynecology, or even subatomic particles zooming randomly out of the void, though *The Book of Telling* includes all these and many more elements. So what is it?

To answer, I'll quote the longtime head of Israel's missile program (and a fabled inventor), who appears in chapter 15. "What does the mind do?" he said. "It searches in a multidimensional space!" This describes invention in a nutshell. The memoir you're holding is concerned with the nature of invention, which is multidimensional—as everyone knows because invention occurs on every level in life. We bring new insights, ideas, and methods to our work, as well as to our family life; we imagine and construct ways of relating to other people or even God; and we constantly improvise on the basic materials of our bodies and characters to invent ourselves anew, in the light of growing experience and changing capacities. Perhaps this process is connected to what biologists tell us: all organisms that perceive, learn, and adapt—which means all living organisms—must change, cleverly and constantly, in order to survive a shifting world. We must change and invent in order to keep on being who we are, an ingenious balancing act on a moving rope. I'd like to think that invention, risky and exciting as it is, is in a literal sense our animating principle.

The Book of Telling is, therefore, a multidimensional book; invention is a theme on many levels of the story. Following are some road signs to the four major levels and types of invention in here.

First, there is the level of technology. I mean, of course, those amazing inventors—the Science Corps and Itzhak Bentov—and their works, both life-cherishing and life-destroying. Shlomo Gur's famous method for raising a prefabricated kibbutz (*homa u'migdal*) overnight is Israeli history, as is the national water pipeline he spearheaded. The Science Corps, which Gur commanded, included the same people who later created Israel's nuclear arsenal and led the country's scientific establishment. Beginning from scratch, by 1949 the Corps had evolved a distinctively Israeli style of technology—an improvisatory style that one Corps engineer, who opened Israel's first oil well, described to me with a half-serious joke. He imagined that Israel had gone to the moon instead of the United States. Now, the Apollo 11 moon shot had required a few trials of manned spaceflights before the actual landing.

"Israelis," said this engineer, "don't like rules. If they'd been up there on those manned flights, I don't believe they would have had the patience for all the steps. They would have said, '*Nu*, let's land already,' and found some way."

I've related how, and under what dire material and spiritual conditions, the Science Corps became the inventors who created a nation's technology and its style. I've also traced Itzhak Bentov's more complicated path: how he became an inventor on his tiny kibbutz, developed under the wing of the Science Corps, and ended up as something he was pleased to call "a Yankee tinkerer," which he pronounced with a heavy Slovak accent. Bentov's remote-controlled cardiac catheter, which he cooked up in his basement, was the first device that allowed doctors to operate in the heart from outside the patient's body rather than manually shoving tubes through blood vessels into the heart. In updated versions from Boston Scientific Company, the catheter is still saving lives today. (A physician I know, who used the old-style manual catheters, recalled how his hands had been inside a patient's heart when the lights went out; he had to hold very still, and he told me, "I sure wish I'd had your dad's catheter then.") Finally, I've written about the inventions themselves. They were, every one, as vividly magical as dreams; they were metaphor, nature, fate. I admired one, was obsessed by another, hated a third; a fourth taught me about vision and illusion—none can be forgotten. A device is the dream of its maker's hands, but its real personality is always unforeseen.

There is the level of history. This book recognizes that Israel's early history is now being reinvented, rather explosively. The afterword and the notes have more to say about that.

There is the level of the personal and the type of creativity called self-invention. I don't like the clichéd term *self-invention*, but I hope to refresh it by meaning very precisely what it says. After my father's death, legal circumstances forced me to prove that, although none of his social circle knew me, I was indeed his daughter and had had a close relationship with him. This experience threw some fundamentals of identity into doubt: Was I a real daughter? What did it mean to be a daughter, my father's daughter? Who was he? Who was I? I had to go find out, put together his story, and likewise reassemble myself. On a much greater scale, others in this book had to do the same—Israel's founding generation, as is well known, built themselves over from scratch according to Zionist ideals in the wake of the Holocaust. A less well-known connection on which this story may shed some light is that between self-invention and invention, in Israel but perhaps also universally. In chapter 13 Shlomo Gur explains what it means to be "No One"—a human cipher for whom, as a sinister blessing, nothing is impossible. He speaks for inventors of all stripes who draw up vision from their own emptiness. In this larger frame of reference, the personal level becomes a cultural one, and personal experience melds with the qualities of culture.

Finally, there is the level of art. The title of this book comes from my having had to engage in artistic invention; I had to create a way to tell this story, and "the Telling" is the result. The Telling, like all devices, is supposed to solve a problem: how do you tell a story belonging to someone else (be it a parent, a friend, an ancestor, or a stranger) as if you had personally experienced it? How do you take someone else's past out of the dead of time and bring it fully alive in your own voice? Each chapter in the book approaches this problem from a different angle.

You may well ask, having read thus far, why it is important to write a multidimensional book. What's the point of this perspective? The point, I think, is what all the book's levels add up to: the mind's search through a multidimensional space or, in other words, the quest of invention itself, which is our human quest to know the universe through individual and collective acts of creation.

Sharona Ben-Tov Muir
Perrysburg, 2007

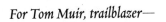

For Tom Muir, trailblazer—

CONTENTS

Part One: A Spark at the Root

One

ON MY ROLLTOP desk sits a paperweight that I took from my father's study after he died, without telling anyone. A clear plastic square, palm-size, it contains what seems to be a wire model of a leafless tree: a silver trunk the thinness of embroidery thread splits into six gossamer limbs, and these, in turn, fray into lines that could be the crow's-feet around my eyes. At first glance, it's hard to imagine how such an exact replica of a tree could have been constructed on this scale. Picking it up, you find a pinhole in the base at the point where the trunk originates; and you discover that the tree is a channel carved by an electric spark. The fine-spun pattern, elaborate to the vanishing point, is what remains of an event lasting a fraction of a second. I often think that the story I want to tell is the tree left by the explosion that killed my father.

At the root of this tree is chance: my father died by chance, and it was purely by accident, years later, that I discovered he had built Israel's first rocket. He had never told me, but then he was secretive. From a chance encounter, I learned that my father had been in a secret group of scientists who had invented weapons during Israel's war of independence; and whose affection for him remained such that, when I tracked them down one memorable summer, they were willing to tell me their story, which is classified and not likely to be fully told for the balance of my lifetime. One by one, my father's secrets were unlocked, after his death, by the same principle of random chance that had killed him; all his mysteries were laid open to that summer's cleansing air and light. Now I've come to a point where I need to tell my father's story as I have slowly pieced it together, from the state secrets of Israel and my own unspoken memories.

Looking out the bay window spanning my workroom, I see the flagstone patio, and upon it, in a black pot, the spindly ruins of a tomato plant. It is mid-September. Beyond, in the sunken garden, a bank of coneflowers, bleached petals fringing their brown nubs. The herb island, raised on rocks from the grass, is a tumulus of lemon balm, mosquitoes, and the dense clustered stars of heath aster. The garden and the trees behind it have a backstage atmosphere, an air of laughing exhaustion; only a few roses are still beautiful, and in our woods, the witchy-limbed blue ashes and stout chinquapin oaks ("sweet oaks," my husband calls them) will soon be as leafless as the silver tree in my paperweight. I catch myself in a loud sigh; half contentment, half disbelief that I can call this house, these woods, home. My first permanent home. My first roots, set down in the early autumn of life, not too late, but barely. . . . I'm thinking of how my father kept my existence so secret that most of his friends were unaware he had a child. And to understand that secret, it is myself I will have to unlock.

Between thumb and forefinger, I pinch the brass pea that is a knob for the little door under the desk's top. Opening it, I remove from the shelf inside a small photograph, deckle-edged.

This was given to me during the fiftieth—but still classified—reunion of the secret scientists. On the back, a note is penciled: "Passover, 1948." I'm translating. The Hebrew reads, *Pesah, Tashah*. Tashah, literally '48, is also the name of a generation. And here they are: the Science Corps, *Hemmed* in the Hebrew acronym.

In the background, a grayish yellowness so abstract it can only be the dunes of Tel Aviv, 1948. *Tel Aviv, Tashah*. The spring of the war. In the foreground, a crowd of twenty or so dressed for the holiday: young women in square-shouldered suits, young men in jackets, no hats, unbuttoned collars spread like wings, and all the wings are white. Each face could be covered by two grains of coarse salt. Yet each has a personality. The mustached gent on the far left conspicuously wears both a hat and an English raincoat, but spurns the camera, bushing his brows. The hip-slung woman on the far right, in a natty suit and ankle-strap pumps, seems to be winding her watch. Midway among these preoccupied college students, smack in the center, kneeling, a crack of light between his thin, baggy-trousered thighs, is my father. Grinning.

When I turn the photo upside down, it confirms that it's not just me; his face really is unique in that crowd. All the other inverted faces look like tiny beads of condensation; roundish, they would jiggle in place if you blew on them. But my father's face is a triangular highlight: he's the drop that's already sliding, committed to motion.

Two years before this snapshot, a few of them had begun meeting clandestinely in cafés, the backs of shops, and their own shabby rented rooms. They called themselves the Association for Security Research, and their goal was to concoct weapons, under a British regime that strictly forbade Jews to possess weapons. One month before declaring Israel's independence, David Ben-Gurion, the future prime minister, turned this handful of lethal dabblers into Hemmed, an official army corps with an R & D budget of $3,000. What, gasped one chemist, will we do with such wealth? He was serious. Hemmed started out in a rooftop shed, improvising its weapons from scrap and guesswork. After the war, the group became a national leadership. Once I had the chance to ask Mr. Shimon Peres, the former prime minister, what he thought of Hemmed. He smiled. It was a security smile. He said they were "an extraordinary, brilliant group. And without doubt, their work is at the foundation of Israel's industrial and defense technology." They gave Israel her national water system, her state institute of defense research, her most visionary tycoon, her Atomic Energy Commission . . .

Except for my father. Who went to the United States to become, first, a self-taught medical inventor nicknamed Invention-a-Minute Ben. Then an author whose book, *Stalking the Wild Pendulum: On the Mechanics of Consciousness,* made him a popular guru, attracting thousands of spiritual seekers. For while his Hemmed friends were struggling to build a Jewish state, my father planned to perfect the evolution of the entire human race, a project whose success, it occurs to me now, might have made the defense of Israel unnecessary. A tall, vividly handsome man, with black upswept brows and a signature laugh, soft and childlike, he lectured to packed halls; and like all true utopians who cannot live with humanity unless they plan to improve it, he was widely loved. Once a woman leapt from her seat shouting "He knows the meaning of life!" So people said, even as my father went insane; his past dismembered his mind as surely as his body would be dismembered by a chance explosion.

. . .

Here they are. Here he is. How thrilling were those meetings with Hemmed, in the summer of 1995! I would stand at the door of some modest Israeli apartment, belonging to a stranger, and press the buzzer under the brass or ceramic nameplate. Once a charming matron flung open the door, and with tears in her eyes placed her soft palms against my face, blocking off portions of it, calling over her shoulder, "Here she is! Here *he* is! Here's Itzik— and *here's* Itzik!"

The thrill for them was Itzhak Bentov brought back to life: his widow's peak, his slightly crossed green eyes in my face. For me, it was the company of those who shared his secret past. Who knew him. I wanted desperately to know my father. I still do.

As I balance the little photograph between my fingers, I feel what can be described as a susurrus wafting up from the picture, tingling my eardrums. It is words: "You must tell this story according to our ancient rules," insists the group, in Lilliputian chorus, faint but penetrating as the cigarette smoke that will wrap them at Passover dinner, at a table laid with the ritual egg, parsley, and lamb shank. They will read aloud from copies of the same book, each guest reciting a section. This book, relating the story of Exodus with commentaries, could be titled *How the Jews Left Egypt,* but it isn't. It is called *The Telling,* as if the act of telling had a separate, overriding importance; as if Pharaonic Egypt were merely one instance of a situation for which the master strategy, in all cases, is telling—and not just any telling, but a kind for which the instructions are specific. In each generation, says the book, each person must tell the story as if he, or she, had left

Egypt personally. It's considered wicked to tell the story using the pronoun "they," instead of "I" and "we."

Now, just how am I supposed to pull this off?

I don't know the first thing about weapons or war. I've lived in Israel, but never served in the army—frankly, loud noises make my hands shake, and I can't even go fishing with my husband because a hooked, flopping bluegill looks to me like a medical emergency. For these doubts, I see, Hemmed does not care. Were they fierce? Did they ask to be stuck in what Israelis immemorially call "the Mess"? The mother who restrains her small son, her hands beside his buttoned collar as he arches gently away from her, did not ask for what's facing her, perhaps has already happened to her. What, after all, did they know, in the beginning, about weapons?

"Nothing," answers the many-voiced whisper, tinged with Misha's comic asperity, Shai's cold judgment, Aia's clear emphasis. "There was darkness in my eyes," whispers Commander No One. Dr. Einstein's advice is a tiny muffled bell: "Always begin with the impossible."

A strange impulse seizes me. Suppose I lived and died in my new home, my first permanent home, without writing another word? Surely it's transience that makes Telling all-important. Only wandering Jews need to worry about remembering events whose traces are left behind, on abandoned continents and in remote millennia. But I'm not a wanderer anymore, I'm home, home for good! Tom and I have found a pile of rocks near our pond, which was quarried out a century ago, when limestone was used to build the coach road through this swampy stretch of Ohio. We've also found rusty iron bolts and marks of blasting in the shelved banks. In 1948, the piled rocks, hauled up from the quarry and dumped beside it, had already sat in our woods for fifty years. From the time Hemmed gathered for Passover until the time I write these words, not a slab in that rubble heap has been disturbed, except slightly, by the nudge of freezing and thawing ice. Why shouldn't I choose to sink wordlessly into that calm? I've never considered it before. But I can—choose. I can bring into the world a new thing, a new Telling; I can be, like my father and his friends, an inventor. Or, thinking twice about Invention, daughter of Necessity, who drove them with whips of war and vision, I can—being American, and feeling the weight of my years—choose to lay down my pen.

The choice hovers before me: a threshold. An open door beyond which lies, by definition, nothing. To step into formlessness and keep going until my joints start to whisper what is not understood, until my heart changes its pronunciation, to be possessed by the opacity of facts, to dream hard, and to invent. Or not.

"Suppose," I say aloud, "I'm just luckier than you were, and we leave it at that?"

My father goes right on grinning. Smiling back, I recall that for him it's Passover Eve, a night devoted not only to telling but also to asking; to raising questions about the ancient story and all its possible meanings until the sun comes up. That brainstorm mood of Passover overtakes me and suddenly I'm asking, "What, exactly, is the meaning of 'luck'? Isn't it chance?"

Oh.

They've got me there, sure enough. It figures, they're physicists.

They've reminded me that chance is at the root of creation. Why? Because of the mysterious way in which subatomic particles can zoom out of the void *at random*. There's empty space, then (briefly violating those laws of conservation you're heard about)—zip!—a speck of matter dances out of nowhere and over the face of nothing, unpredicted, unforeseeable. This is matter, the stuff of our bodies, of stars and planets, of all that lives and all that is. And it is conditional upon chance. The great mythic World Tree with its leviathan boughs and superclustered leaves rises from a random spark; by luck, you might say. I'd thought that my luck in being an American, in a safe home, separated me forever from the little crowd posed on the drab, windy dunes of their dangerous world. But my luck has unveiled its nature, and I find that it doesn't stand between me and anyone, or anything. It is the exact same size as the universe.

Now as I look at Hemmed, every face looks back at me. They're huddled, waiting for me to commence writing about the Hollow Charge, the wooden Shark, the Automatic Dog, and my father's Loretta, as if I, personally, had made them. Where do I begin? With whose voice, whose face? At which door in memory do I knock first?

Let it be the varnished pine door with a round brass doorknob, looming higher than any door I've seen since childhood. It is the door to my father's basement laboratory, and this afternoon—it's standing open.

Two

"HOW—?" I ASKED.

My father fished car keys from his khaki shorts pocket; he was in the rare mood that forced him out of his basement laboratory, where fluorescent lights burned even on a bright summer day.

"Let's go, *yalla!*"

I rubbed the small plastic disk one more time. My fingers told me plastic, but my eyes told me a chess piece, the ivory arch of the horse head tinged by a green sourceless illumination. Obediently, I replaced the disk on the coffee table that smelled of oiled teak, in the living room where the chairs and couch, hemp rug, and linen drapes were all shades of beige and brown. These were the colors my father said did not scream. The bronze statuette that my father's wife had made pricked my heart: a leafless bough encircling a child perched on it, one leg awry. How could she suffer from the same lonely feeling that she gave me?

"*Yalla*, the canoodles will be gone."

"You made us late, and all that is left is the lousy aluminyoom canoe instead of the nice quiet fiberglass."

But his grumble from the prow wasn't serious, more of a greedy croon aimed at the Concord River, its current paisleyed with bright green algae, the tunnel of the trees surrounding us with shafts of dusky light in which dragonflies purred on sheer wings. Anyway, I liked the watery *clock clock* that came from the gliding shadow of our metal hull.

"Daddy, a turtle." We drifted toward the slate and granite stones, where one stone had just stopped crawling.

"He looks to see if you are interested in eating him. He is lazy and fat." The turtle craned its scrawny neck in an insulted way.

"Fatso, fatso," my father clucked at it, beaming.

"Look at his red stripe," I whispered.

"Now he is thinking better of the situation—hopla! There he goes." The canoe slid back into midstream, my father murmuring, "If you are a hungry raccoon you are disappointed now, Pookli—no turtle soup, no turtle gloop, you are unilaterally turtle-less. . . ."

I sobbed with giggles, bobbing on the seat.

"Oh, say it again!" My paddle swooped out of the water and fell with a liquid crash.

"You scare all the animals!" A fierce glance over the wedge of his shoulders, dark amber beneath the white mesh shirt. Quietly, we paddled deeper and deeper into the humming and bubbling summer hush. My father pulled us along strongly, his bare arms working little except under the skin. Sunburn began peeling his bald dome, above the rear ruff. Now and then, the outside edge of his glasses sparkled.

"*Nu,* what'll we sing?" he called in Hebrew, without looking around or interrupting his stroke.

"*Zamar Nodeyd?*" We la-lahed a bit to set the key. His voice dropped octaves when he rumbled out the first words, "The way is long and great, and great its splendor." My harmony sounded like a balloon on a string that rose and dipped to the rhythm he set. "All walk the road foreseen and bitter. But I stride alone, the wandering singer."

This duet was our favorite because it lasted a long time, as the man in the song poked fun at the other men who strove for gold or settled for love; and said he didn't need a house with a green field, just an echo for his song; and wandered through several verses. "The Wandering Singer" was our antidote to traffic jams, and it had a fun refrain.

"*Ha*—leiloo," I pealed.

"Hallei—*loo!*" he boomed. We burst out laughing. Then we tried a pioneer song, but that didn't go so well; after hectic praise of wine vats and grain bins, fertile cows and homes overflowing with babies, I let my father down in the refrain. He wiped a handkerchief over his shiny burned head, and the canoe listed; the bilgewater around my bare feet hunted for an exit.

"You have to pour yourself out—you didn't pour. Try again."

"What *more* could you ask of us, beloved land, that is not yet?"

"Tsk. What—*more*—could you ask! Of us, beloved land! Never mind. Time for the island."

Why couldn't I pour myself out? I felt puny, like an eyedropper.

Traveling upstream, we were slithering through river weeds that swelled like waves and waves of loose hair just below the surface. Rushes hid each bend, their stems massed in mud bars. My father could find the way that the river itself seemed to have lost. All I had to do was stroke, drip weeds off the stern, and wait. I planned to ask him about the plastic disk as soon as we tied up the canoe and unscrewed our bottles of orange juice.

The granite ledge where we moored was like a private deck;

we'd never seen other people or signs of picnics there, even though the islet, a tangle of dwarf pine and raspberry cane, faced a widening-out of the river where it ran deeper and blue.

The trick was to catch him before he fell asleep.

"You promised to explain," I nudged.

"Yah!" Sprawled and leaning on one elbow, he picked up a couple of pebbles and flicked them off the bank. "I told you what an interference pattern is."

"That"—I pointed to the two rings whose widening ripples crisscrossed, like curved tic-tac-toe, and were fast diminishing.

"Good little Pookli." He pinched my nose and wiggled it, not with infuriating condescension to an eight-year-old, but as if my nose were part of what he had to say. "Now, such a pattern can be made of waves of water, or light waves. Or it can be light waves from a ruby laser. How is that different?"

"I don't know," I fretted, itching with suspense.

"You do too know, you are lazy and refusing to think."

Wiping his glasses with a fresh handkerchief, he returned my glare, stony-eyed as a bird. The bridge of his nose looked pale and used. I made an effort, and "ruby laser" struck my mind with a fearsome, magnificent wave of red, a trough of black, another wave of red, like a chain of identical mountains marching. Every one in step. Like my heartbeat through the pillow, at night. He'd given me a name for this.

"Coherent light," I remembered.

"Right! *Nu,* good. The disk is a thing called a hologram. We make it by splitting a laser ray. One half bounces off our chess piece, while the other half goes about his business as usual. Then the rays meet again, on a piece of film. The one who's been out in the world bouncing off things is now different from his brother;

he's a sophisticated ray with fancy ideas. So when the two rays are reunited, they get into a big argument which is . . . ?"

"Interference," I yelled, startling myself and some ducks, which took off squabbling.

"When we shine a laser through the film, a 3-D image appears in midair. You saw a kind of photograph of that image. Pretty impressive, hey? Like the magic in your hobbit books?"

"Can I have it?"

"No, it belongs to MIT. But listen to this!" He went on in a low, sorcerous voice that turned a passing motor boat's drone and the faint spiel of a radio into nothing.

"I take the hologram, shine light through a spot on the edge of the film, and can't believe my eyes. What is appearing in thin air? The white knight! I shine it on another spot. The white knight again! Why? Because it seems, little Pookli, that a part of the interference pattern contains the same information as the whole. Even a tiny bit on the edge contains the whole image, horse and rider together, as the Bible says! So your daddy is thinking a lot about this phenomena and what does it suggest?"

"Phenomenon," I automatically fretted.

He brought out the punch line, with three pats in front of a stunned, feeler-waving black ant who'd been surveying a granite scale.

"The universe, works, like a hologram."

My face bunched. My back froze. I didn't understand, yet I did. Like an iridescent feathered neck, a wingtip—gone! Stillness . . . Then I caught it.

"We can see all of it, in a part of it?"

My father nodded as though he'd shared the very secret of delight.

"How?"

"Must work out the mechanics." He paused, the air around him bright with his gathering thoughts. "We live in a vibrating world. Everywhere are waves. Sound waves, light waves, electro-magnetism, the ballistic energy of the body—I mean, Pookli, the way you are subtly bouncing up and down to your heartbeat, and your brain bounces in your skull too"—he grabbed my head and ruffled it—"like a little peach in a can. And its little thoughts produce an electric ripple that spreads all the way to the stars. In fact, this whole world is rippling, shaking like jelly." He rolled onto his back, and watched his steepled fingertips play at typing. "You know, if you hang up clocks with pendulums of the same length, they eventually swing together? Because the vibrations they transmit through the wall align them. Or women's—*nu*. My hunch is, all these wave energies, affecting one another, make a giant interference pattern containing all the information in the universe. The mind uses this information. How, though—your daddy must figure out! You, too, know it, intuitively. That poem you made up—time something-or-other?"

I returned abruptly to myself, to grubby kneecaps, to my short-haired, bespectacled, impossibility of fairy grace. He'd made me remember how our weekly visit had begun that morning.

I had waited outside my mother's apartment, on the landing, in the odor of old banister varnish and rubber mats, while my parents fought. My mother in her horsey way, with stamping and neighs of rage, my father with his all-conquering laugh. Two Hebrew words shot out of the doorway.

"*Kessef!*" Money!

"*Meshuga'at!*" Madwoman!

"Oh, you can't afford it," my mother ripped into English, "but you can afford to build a studio for that goddamned Russian—princess!"

A dead silence. Then my mother's bray made the banister I held seem to vibrate. The round nostrils at the end of her perfect nose would be black, and her mouth stretched like opera.

"Laugh laugh laugh laugh! Sacrifice us all to Sonia from Kiev! I'm crazy, am I! Not her! I've been told what goes on in your lovely home! Why, oh why! Why!"

The powerful mare turned into a woman whose every breath was a wail. Her neck would hang from the ivory knob at her nape. My father and I would go now.

But in a high, weird tone he never used with me, he said, "Her family saved Jews."

"Oh really," my mother replied, as if restored to herself by knowing better. I heard the floorboards creak. My father and I would go now.

It had all happened because of my lousy line that wasn't even a poem.

"I forget," I muttered.

"Something about time, rhyme, chime, what?"

It was unbearable when words were thrown around without respect. I stared at the rock between my shorts and bare calves.

"'Time heals all wounds but the wound of time,'" I recited. Mucus hanging from my lips during the flu had sickened me less.

"And your mummy said these words came to you from a voice."

I brooded. Shouldn't have told. But I'd heard the words spoken aloud while I was snug in bed, in the sunlight; when I told my mother she'd hugged me and cried, "Angel girl!" And we'd eaten cinammon rolls for breakfast.

"Do you know where the voice came from? Pookli, you are not listening. It existed somewhere in the universe and was waiting for you to use it. I too hear voices when I am working on an invention. I sweat sweat sweat over the non-clotting hypodermic needle. One day, out of the blue, a voice says, 'Hey there, if you want to find out about non-clotting design, why don't you visit a serpentarium and see how the snakes that make their living from piercing animals solve the problem?' Hee hee! Eh, Pookli?"

He blinked comically till I laughed politely. "This you have from me. All creative people have it. Which is also why your mummy is silly to scream at your poor daddy about sending you to private schools."

He tucked his arms under his head. He'd put a shirt on, and the damp undersides of his sleeves were a bit pitiful to see, like a snail's foot. I looked away. A motorboat sputtered past, steered by a bare-bellied man in a Red Sox cap; the wake knocked under our moored canoe. The day was awash in blue and green smells, a gorgeous day, everything glad and pouring itself out. Only I was bothered with questions that put me in a nervous, tight, eyedropper kind of mood. I could live with unanswered questions but not with unasked ones.

"But Daddy, there aren't any waves in outer space."

"Ridiculous."

A line shone between his lids; it was how he kept track of time during a nap.

"Why?" He acted asleep. I puffed my cheeks and drummed on my knees.

"But it's vacuum, Daddy."

"Yakyakyak."

"But you *said*!" I hurled a stone. My father blinked at the plop.

"What is this throwing going on?"

"Daddy, please, just tell me."

"Has the sun stopped working?"

"I forgot about light waves!" My father chuckled.

"If the Almighty forgot like you, *oy va'avoy.*"

"Okay, but . . . everything's made of atoms and they don't shake."

"And how. Atoms vibrate like mad."

"They do?" I sighed, and searched for a mental picture of an atom that shook; a spiderweb with trembling raindrops, maybe. That's how I'd remember.

"Pookli, the universe is full of waves. End of story. Now be quiet so I can rest."

"Please can I ask just one more thing?" I stood up with my hands clasped so he'd feel me waiting. He groaned and crooked an index finger.

"How do you know the voice isn't just a dream?"

"You yourself know," he muttered, and snored before I could respond. The answer was immensely reassuring. I got down on all fours and backed off the ledge into the river, enveloped to my knees in the coolest and most delectable-smelling water that tickled as I waded, looking for the pink roots my father objected to; they stuck rustily onto the seat of his car, after I'd left. Silt and soft pine needles clouded my toes, greenish under water like the color of the hologram. I looked around at the world my father's words had changed. Waves swung gently against my legs. Across the river, dense leafy trees bowed to the changed wind. Sun pulsed against my closed eyelids, and inside the red, heartbeats were traveling. The waves of the world carried a feeling of deep trust to my lips, and I smiled. Then my balance faltered, eyes opened, and goose bumps covered my wet skin.

I climbed back onto the ledge, careful not to touch my napping father. The three territories of his body—the head between elbow peaks, his chest's arched plateau, the outflung marches of his legs—did not stir to wakefulness. Maybe today Daddy would forget the time. We'd stay here and build a log cabin. I would go barefoot always, and learn to fish.

My memories of my father are a small collection, but exceedingly precious, that I've carried everywhere I've gone. They are my jewels. But I wonder. I wonder. We'd been canoing on the Concord River. I remember the smells, that summer evening when he drove me back to my mother's apartment. I smelled of pond water; the car of heated vinyl; my father of his stubborn acid fustiness, as if he'd changed the combinations of human skin, down in the lab. I remember the warm tar undertone off the parking lot, and a rotten whiff of the garbage cans that stood at the apartment building's back door. My father's car, a ponderous Plymouth Satellite, glided into the parking lot, signaling *glink-glink*. On the maple trees, the leaves looked like ragged velvet. Through their canopy, the sky was filled with tempered light that would not dim noticeably till late. It was an hour soft with summer melancholy; an hour when, as an adult, I would open a book and a cold beer. But then, of course, I was not in command of my resources.

In the parked car, my father consulted his map of greater Boston to review the way we'd come, something he always felt he had to do; then folded the map, leaned over, and slipped it into the glove compartment. He kissed my cheek, told me to be good, and turned the keys in the ignition. I had no idea that I wasn't going to open the door and climb out. The decision came from my legs. I looked at them; they looked like they felt. Boiled macaroni with knees.

"Pookli, you go home now," my father prompted. I said okay and waited for the mood of my legs to change. During that delay, my legs proposed that I sit tight; they hinted, further, that I was curious to find out what would happen.

"I don't want to go home."

I was stunned at the possibilities that tumbled through the door of these words I had said. Now we might drive away together. He leaned across my lap, a big beshirted father-smelling wave, and pushed the chrome door handle. *Glunch.* The door opened just a crack.

"You go on up to your mummy."

"She's mad at me." I meant, she's mad at you: parents forced one to speak in circles. I began to see my mother's mood in the fire escape going up and down the building, black cage black cage black cage. The back stairs had three times as many flights and landings as the front, and were narrower, harder to catch your breath in. There would be a thick silence as I entered the apartment. Then questions. Was Sonia in the house? What did she say? What do you mean. Then I'd creep away dogged by her cry, *Just like your father!* Which made no sense, since my father was a man, free and safe, with a big house. I would go to my little bedroom and close the door, and as I sat on the bed in a vigilant pose I felt that the walls were made of nothing but tears.

"Yes," my father agreed, "your mummy likes to get mad. Terrible."

"I don't want to go."

"Women and their moods," sneered my father—who never said those things. It was my mother's father, my grandfather, who insisted that women were stupid, dirty, and made out of leftovers. Now because of what I'd done, the rules had changed.

I can still feel the logic meshing trickily in my half-grown brain, working to undo my father's bad opinion. Without words, roughly like this: I must prove that women's faults were not my faults. Or, that women were okay. Which? I wasn't a woman, yet. This I would not use. Though it counted—it was why Grandpa talked baby talk to me; but there was something feeble, shamed, about being okay that way. Besides, it was temporary. My father would not accept a slipshod reason. The truth had to be good enough. But what was it? What was it?

"Not all women have moods," I ventured.

My father's expression was unreadable. Then he looked away, through the windshield, at the planet covered with women, and said, "They all, all have moods."

"No they don't," I bluffed. He couldn't prove it! I had won.

"Now, Pookli," my father said in a kindly voice that raised my chin right off my chest. "You want to be a doctor?"

I was so glad he remembered.

"You will become a doctor, little Pookli," he promised, and my soul turned over. I would ride away with my father to live in his big house and become a doctor, and I would remember, always remember, how I had done this by saying what I thought.

"You will be a psychiatrist," he said. "You will be the best psychiatrist in the world, because with your mummy, you are learning all about crazy people. Eh? Crrray-zee pipple," he hummed, making his clown face, expecting me to laugh.

. . .

Next should come a sentence like "Through the open car door, I saw my sneaker toe reaching for the asphalt"; but that's a lie. I don't know what I did, what I felt, or who I was. I remember nothing. It's as if a switch were thrown. The child, her bare leg, her thoughts, tremble and vanish. . . . From the last moment I remember, it seems that I too, like the laser ray that creates the hologram, am split into disjunct selves—a woman who cannot find the child she grew from, a child lost in the quarantine of an unknowable past. I don't know whom I have forgotten, with her, or who I might be, if I remembered.

But suppose.

Suppose she materializes elsewhere, in a dim room, perched on the edge of a brocade armchair, legs dangling above its polished bombé feet, in the blown hush of an electric fan planted on the floor. The plaster walls, aged to amber, are hung with framed paintings at one and the same level: winding copper rivers, the sea, a fluid dancer in pastel blue. Bodies of water. The child feels as if a canal of imaginary water surrounds her, murmuring, as she waits. She waits for thirty years. All that time she cannot remember her name, who she is supposed to be, exactly as in a dream; but the years that are passing are not dream time, they are her life. Until Commander No One enters carrying a pitcher of water and a glass, on a tray from whose rim the silverplate has worn away.

Three

I ARRIVED IN ISRAEL in May 1995. I'd rented an apartment, sight unseen, in a Jerusalem neighborhood where I'd once lived with my mother. I'd often been back, but my first walk through the city always made me smile. I climbed the steep streets overgrown with sky-blue plumbago, pulling my rolling duffel behind me and breathing in the air, like dry champagne, that was scented with conifer resin, jasmine, diesel, and dust. Spring had brought feral kittens to the sidewalks—a peach-sized tabby raced past, attacked by a dark puff with ears. New construction had gone up, faced with rough limestone that caught the light in fleshy facets. Bougainvilleas engulfed some balconies like a migration of magenta butterflies, from which emerged the peak of a canvas umbrella, or a rail hung with potted geraniums, or an entire family eating dinner, cutlery clinking. The trashmen were on strike again, and on some corners pastel-colored trash bags attracted flies to their opalescent heaps; what else is new, I thought cheerfully.

Look at all these ATMs on Aza Street. Such a nice boutique for fancy straw hats. Such an elegant little florist's, buckets of gerberas on the threshold. The same grocery! I couldn't wait to smell the yogurty fragrance of that small grocery whose proprietor was hunched by the radio, intent on the soccer news, and who would perform my Israeli miracle. Since I'd landed, face after face had turned to me without the guarded flicker I endured in the States, where my looks caused mistrust; I'd been told that I resembled "a tigress." I was five foot ten with bushy hair and too-intense eyes, and the worst of it was, none of this matched my timid personality. But here tigress women ruled the home and nobody thought twice about it. The grocery owner (whose eyes would also be too hot and slightly crossed) would count out my change with a chivalrous air, deferentially, for a soft-spoken American. In the U.S. I might be a big bad broad. In Israel I was a lady.

Ah! Here was Yom Tov's! I paused on the curb. Yom Tov's green awning covered a sunken courtyard where fruit-stacked and vegetable-filled bins, laden with tints and musks, shed a glow whose effect on the senses, even at a distance, was profound. Cardboard signs, stuck drunkenly atop the bins, carried high prices beside apt quotes from scripture. I smiled like a fool, shoved on both sides by hurrying passersby. Here was where I used to get off the bus from school, the #9, an endlessly bombed line, but I wasn't remembering that. I was remembering the man in the blue stocking cap from Yom Tov's, who would meet the kids getting off here with a tin tray of peeled cactus pears, ruby-ripe, in dirty water. A mouthful of pits like shot and pulp like chilled May wine. I was twelve . . . and now, I rolled my rattling duffel around the corner of Arlozorov Street, to my new place.

A hallway, cool and dark, led toward the sunlit lace curtains stirring against wrought-iron balcony doors. Parking my bag, I walked down the hall lined with the books of my landlord, a historian, pausing to hide my passport between *Theories of History* and *In Her Majesty's Secret Service*. The hall ended in a friendly worn-out sofa and a TV. I turned left, and looked with joy around the kitchen. It was a big gray kitchen circumscribed by an old-fashioned marble countertop. Over the sink arched a chrome faucet of the old style, with its propensity for bashing one's brow. And a little gas stove by the window. The Turkish refrigerator, tugged open, showed that my kind landlord had left a dozen eggs and a bag of milk. I located the only fixture not dating from the sixties, an electric kettle with a lime-caked coil. I set it to burbling, and, hands on hips, fluently ordered the kitchen to reveal its rusty can of Elite. The kitchen, seeing who was who and what was what, complied. Pleased beyond measure, I sipped my instant coffee at the aluminum table, reading the note left under the telephone. Four plumbers were listed in order of despair. I toyed with the phone. There were scientists to call, a story to test. It was a touch-tone phone with round rubber feet, whose receiver could not be banged down, only lightly clicked. This was an instrument of the unknown.

What brought me here was the accident of meeting Edo Peled, who looks precisely like a young man in my Passover photograph of Hemmed. This is Edo's father. His head perches between a woman's hair and a man's shoulder, radiating strength from the level eyes, the straight porch of his mustache. His body is hidden but I can imagine it, since Edo was my lover—in a luckier, nonexistent reality, where both our fathers lived out their natural life spans. For like mine, Edo's father died betimes, in the U.S., in fire. What I know about him is this. In Hemmed, he developed the infrared lanterns used in an attack that sank the Egyptian navy flagship *Emir Farouk*. The attackers were five men and three Italian pigs, but that's another story. After the war, he was a leading figure in the creation of Israel's nuclear reactor. And he had a nickname. All the Tashah generation had them: Yevgeny was "Genka," Yonatan was "Yoyo," a Hemmed woman was, quite innocently, "Hollow Charge," and a Hemmed man with no feminine qualities whatsoever was "Lola." My inquiries about Edo's father drew blanks: "Peled? I don't remember any Peled." Then light dawned. "You mean Plafchik! Why didn't you say so?" That's it, for Plafchik. I can write more about his son, whom I never so much as kissed, but who sent me to Israel to look for Hemmed.

I'd been a graduate student at Stanford. A gay friend of mine with universal matchmaking tendencies had advertised for a housemate, and called me to report the result.

"He's straight. But he's very nice. And he wants to meet you, Sharona."

"Oh shush."

"No no no no. He knew your father."

"What?"

"He met your father and your—stepmother or whatever, and he wants to meet you. Really, you should. He's nice. Very honest and burly."

"Are you sure?"

"Well, I'm not *liable*."

"I mean about meeting my father." It was inconceivable that my father's sphere should casually intersect with mine. But it happened. I was perched on a low limestone wall in the afternoon sun, while Edo sprawled beside me, wolfing a ham sandwich and an apple. We spoke with the directness of Israelis and the immense constraint of lonely children.

"It bothers me how it smells like the Land here," Edo pointed out. "From eucalyptus and . . . dryness, and . . . but it's not quite right, it's like . . ."

"Sanitized. The Land after the dry cleaner's."

He had a rich pleasurable chuckle. Before us stretched the Oval, a green lawn where near-nude students flipped onto their bellies, reading, or onto their backs, holding books over their faces. Edo told me that as an eight-year-old boy with his parents, visiting from Israel, he'd seen my father's house in Boston. He'd been promised a little girl to play with, but there was no little girl, and the atmosphere had made him uneasy.

"Sonia had this tooth, you remember? Her tooth? It stuck into her lip and lipstick was smeared on it. She made me think of a vampire." I shrugged. Fastidiously, he brushed crumbs off his khaki shorts, adding in his baritone, "My father really liked your father. You know he built Israel's first rocket."

"Your dad? Wow."

"No, yours. Your father's rocket was the first."

I shook my head.

"You must be thinking of a different Bentov."

"No, that's how my father knew him. They were together in '48. Didn't you know that? Didn't he tell you about Hemmed?" And my bafflement. And Edo's voice swelling with factuality.

The thing about Edo was, like me, he was at odds with his body. The slab of his chest, the trunks of his arms. Knees like sanded burls. The globes in his calves. The great feet. He took up plenty of space, like well-made furniture. Even his dark, intelligent eyeballs were massive. His hair was wool. His watchband snagged in it. When he went to the Israeli folk dances at the Jewish community center, mothers of eligible daughters flew to him, because it was impossible not to see three, four, five toddlers clambering up those knees, riding those thighs, banging on those biceps. Edo was born to be a father and a play set for pure-bred Jewish children. When the girls danced over to him, he felt himself a marked man. Which was not, in itself, a bad thing. But he had a mother and too many sisters. Well, only two sisters, but the foibles of his family—of which he was now the head and prop— tended to magnify them. Mrs. Peled was one of those lovely, maddening souls who do not take the business of life seriously. She met me with tearful kisses, but sadness had no grip on this matron who dressed like a kibbutz girl of the forties, her gray braid over her shoulder, eyes alight. Her true passion was animals. Back home in Rehovot, chickens, vultures, goats, a donkey, a pony, an injured stork, dogs, ducks, rabbits, a coyote pup, and incessant cats (but no pigs, Italian or otherwise) came and went in her courtyard, and colored her view of human affairs, I'd guess. She may have felt that any animal possessing speech was well provided for, and needed no more than her vague affection. Now the sisters were pursuing an erratic course of veterinary studies, running out of funds in Romania, Mexico, Kenya; their brother accepted collect calls at weird hours, handled crisis after crisis, rescuing Talia and Amalia and underwriting their latest miracle sheep dip or prizewinning saluki. Edo was a doctoral student in computer science, with a logical mind and a quiet, delicate character. He couldn't help it if partnership with the opposite sex appeared to him as an unrestful affair, zoomorphic and frighteningly expensive.

We might have kissed. We were good friends, fantasizing and complaining in each other's company. We used to go to the supermarket, checking out the liquor aisle for the latest California wines. After sampling them with me, Edo would drive me home. Sometimes there was a moment at the curb; our bodies blurred, our eyes kissed. But we were too much alike, I suppose. We both had the habits of the chronically unsettled. After my father died, I'd begun hiding my jewelry under the torn lining of my handbag, which hung on my shoulder wherever I went. Edo kidded me about the thunk my handbag made on the floor. I teased him about his collection of junk cars, which he didn't keep in the usual way, in his garage or backyard. He parked them all around Palo Alto in an infinite series of backups, if you understood how minds like ours worked. You had to imagine him leaping into a strategically placed jalopy and roaring and clanking out of town.

After graduation, Edo moved to London and married an Irish Catholic veterinarian who specialized in the health of hogs. Before leaving, he wrote a name and phone number in my address book. Make the call, he urged. I was afraid to bother someone as important as Professor Gil, one of the handful of pioneers of Israel's nuclear science. Edo looked at me. "He was your father's close friend, maybe his best friend," he said.

More than a decade passed before I made the call. It was a step, for me, perhaps a bit like marriage for Edo. Our fathers had imposed, on both of us, burdens that blocked certain steps in life. My burden involved a missing pearl necklace. What did a string of pearls have to do with not calling an Israeli physicist? Please don't ask me to explain just yet. I can say that the reason I finally made the call, in a nutshell, is the same reason Edo had for finally getting married: I ran to the end of my tether. And at the end, I found a dream.

I had this dream. It was night. I was in the office of a university physics department, alone among the fluorescent lights, shut-down photocopiers, secretaries' potted plants. The office had glass walls; I looked through them into a small cobbled court, in the center of which stood a luminous pillar of crystal. On the pillar's summit, higher than the tops of the dark trees, a boulder as huge as a ship was balanced, blazing white. I understood, then, that the crystal pillar was a moonbeam, and the looming boulder the moon, but seen from their own perspective, as they experienced themselves. I was awed by the full moon whose effulgence solidified into a beam of conscious will. Looking down, my gaze guided down the clear beam, I saw a wooden bench. A man sat there, chin sunk into his coat collar. He'd been waiting a long time, in the cold and the dark. He waited frugally, without great hope—with strictly enough—for me to come out from behind the office's safe glass walls, because he had a message to communicate about the nature of reality.

I didn't need to identify him. The dream was what he was.

This dream, in the winter of '95, sent me in pursuit of Hemmed at last, applying for travel grants, booking flights, packing bags. I believed with absolute conviction that Hemmed would give me the vision I needed to revive my life. Because that was the tether to whose end I had run: a lack of vision that was like suffocating slowly.

It began some years after my father's death, around the time I failed to win a lawsuit, extravagantly painful and expensive, for wrongful death. It began although I had just sent a collection of poems to an editor at Harper & Row "cold"—that is, without contacts or reputation—and she'd liked it. The book was published in 1985 (around the time I met Edo). It seemed that my dream for many years, to be a poet—the only job I did with sincerity—had finally been realized. I was feted: I read at a banquet of *The Paris Review,* and Mr. George Plimpton, observing that I'd polished off my shiitake mushrooms, fed me from his own august plate. Meanwhile, the second worst factor in my history was turning into an undeniable fact. I had dried up. It was not writer's block. Every day, I clattered away for the allotted hours at a chunky, encouragingly humming electric typewriter that smelled of overheated ink ribbons. What emerged from my industry was not poetry at all, ninety-nine times out of a hundred, but a verbal gurry that made neither sense nor good nonsense; if it resembled anything, it would be the phrases, mildly grotesque, that are uttered aloud in sleep. Any fresh subject that I touched was worked down into chaos. I couldn't prevent it while I was writing; then, every line glowed with beauty. On a second reading, even a few minutes later, the same lines were incomprehensible. Gold kept turning into sand, the fast-running, hexed sand of wasted promise.

Hell, at least, had meaning, inspired poetry, *lasciate ogni speranza.* Where my mind went, when I tried to write, made hell look comparatively cozy. It was a lack of . . . no, more like a void. The opposite of vision? The negative of a negative that did not make any kind of positive? Missing, still, the right words for it, sounding awkward to myself, I wonder about the writing that my mind sent back from its darkness. Those hundreds of non-poems and un-poems, my great, worthless opus. Suppose it were like a hollow rubber ball that could be flipped inside out, into an everted perspective, from which might suddenly emerge beautiful truths? For in some such faith it was composed—I don't say that I composed it. We don't know our minds, we do not know the inscrutable process called mind; and how should I presume to write the word "I" as if that were my mind's name? Perhaps it was another name that my mind was composing, all those years, when I would get up from my typewriter with cold muscles and a blasted heart, as if I'd done something, or been somewhere, unnatural. I knew I'd gone grossly wrong, but no mental illness offered its symptoms as an explanation.

While this state dragged on, every day felt alike, whether good, bad, or usual. My days felt like prostheses, substitutes for the real days that had gone missing. Often I had to suppress the insane question of whether I was myself, or just an illusion who felt like me, but wasn't. Or I imagined that I'd slipped into one of those "other realities" my father had talked about; I had gotten lost and was stranded forever in an alien dimension. When this thought crept up, I cringed: my father's madness lay that way. Besides, hadn't I sworn off his sci-fi worldview after his death? Eventually I formed the disheartening idea that my heredity had required a trade: to stay sane, to live without my father's demons, I had had to lose my creative powers. Visionary madness, or sterile sanity—somehow, I'd picked the latter. That was the tether.

At its end, I was thirty-nine years old.

Then I had my moon dream, which felt, upon waking, like the signal answering an SOS.

Pondering the telephone, I was startled to notice that the door to the balcony stood slightly ajar. A grapevine had climbed through the latch mechanism and twined in the wrought iron's floral patterns. The iron loops were leaves in the abstract, filled with ruffling, sunstruck layers of real leaves. It seemed an auspicious sign for my next step.

The voice, on the other end of the line, was a little boy's.

"Is Professor Gil there?"

"Which one?"

"Professor Gil the physicist?"

"Which one?" Children who answered phones. I tried again, I hoped, calmly.

"Please, I'd like to speak to Professor Uri Gil."

"He isn't home. You can talk to Professor Sara Gil." He shouted, off-receiver, "Gramma!" Another line clicked on.

"Yes?" inquired a tired voice.

"Professor Gil," I began nervously, "my name is Sharona. I think you knew my father, Itzhak Bentov."

"Sharona." The voice was transformed. "Your father was our dear friend. Welcome. It's miraculous that we're hearing from you."

"Thanks," I said.

"We tried to contact you after Itzik's tragic death. We called Sonia and asked for your address, but she said she didn't have it. Where were you?" I set my warm coffee mug down on the table, and slid cross-legged to the floor. I chose between the truth, which would be awkward, and something truthful to its side.

"Moving around a lot," I said.

Sara Gil invited me to dinner on Friday of the following week. When I asked, feeling silly, whether my father had built a rocket, she said straight out, "Israel's first rocket." My father's assistant on the rocket lived in her building, and she planned to invite him too. Crowning all this, Friday is the Sabbath, Shabbat, devoted to family. I was to have Shabbat dinner with my father's friends. This had never happened before, and my first reaction was fear—of disappointing the Gils. I imagined the two elite intellectuals saying to each other, "She's an ordinary girl, not up to her father's standard." Then they would lose interest in me and my search. For two days I shopped and fussed. I bought the most expensive chocolates at Avi's American Espresso Bar; Avi himself fixed me a macchiato on the house. I bought stylish

34

Galilee wines, red and white. I polished my best pumps, buffed the handbag, and ordered an air-conditioned taxi for the drive from Jerusalem to Rehovot, so as to arrive fresh.

Then the externals were ready but I was not. My father's assistant! How to prepare for this meeting? Rocket scientists! Men in white lab coats, I pictured, which contradicted everything I knew about early Israel. For instance, tents. They were furnished with vegetable crates, for cupboards and chairs, brick-and-board bookshelves, and an oil lamp hanging from the central pole. "The best house," my father used to say, "is the one you fold up and carry with you." Definitely not white lab coats. What? I longed to know. It'd be like getting a new memory of my father, instead of the old ones I knew by heart. It would be like bringing him back to life.

So I went to visit the old British police station and prison, now called the Museum of the Underground, where I had never gone. Few did. Israel is not India; reminders of the empire are scant, and somehow don't remind. You walk down King George Street without thinking about it. But climbing uphill past the stone-faced blocks of government buildings, the Central Post Office, Internal Affairs, flights of steps or stretches of asphalt making me pause for breath, I did remember that the colonial administration had sat here, surrounded by rolled barbed wire, on this urban hilltop. "Government Hill," right. But everyone had called it Bevingrad, after old Ernie Bevin. Huh.

The museum was a rough-hewn Ottoman house. It stood at the back of a bright ocher courtyard, separated from the street by a thin comb of pine trees, a bench or two, a crow, and a deep interior peace, as if the past really dwelled here, unchanging and complete. Inside the museum, the reception area was empty. A black manual typewriter with a white enamel emblem, IMPERIAL, sat by an open logbook, on a wooden desk that a Punjabi clerk might have just left. Softly, I read aloud the faded roster of names. Then the curator entered, plump, dowdy, full of personal smiles; so I had to tell her about my dad in the Palmah. She too had been in the Palmah. Her war was this empty museum, now.

I was drawn to the display in the prison's main corridor, walking there alone under the peeling arches. A collection of newspaper columns was posted on the wall, announcing executions. I looked hard at the faces of the condemned, whose enlarged black-and-white contours spoke of nothing but their youth. Shadows barred a square of sun fallen from the skylight to the dark floor, as I peered into a cell large enough for ten prisoners, which had held thirty. Woven rag pallets lay in rows. Tattered books on a plank. One, with a hand-embroidered cover of black cloth and white stars, was poetry, *Heaven's Distant Stars*. A tin cup hung by its rope handle, from a dry faucet, and the water it had not held for fifty years seemed to start from my eyes. Here was the danger my father had faced; his life might have ended here. Everything—everything I had thought of as "history"—was real.

The night before meeting the Gils, in the small hours, I stepped barefoot onto my bedewed balcony and leaned, blinking, into the chill indigo night. I heard a baby wail, and smelled an angelic sweetness drifting from the stars, or from a nearby bakery. I felt as excited as a child.

Sara Gil was slightly put off by the wine and chocolate.

"There was no necessity for this," she said as I handed her bag after bag, in the foyer of the sunny apartment.

"Wine is always a necessity," I joked, earning a keen look. With graceful reserve, she led me inside. Their living room, which Israelis call the salon, was furnished in the Danish modern style that my father had liked: he'd owned this same flat brown sofa where I sat hastily arranging my pleated skirt while Sara went to fetch Uri. Their taste was shared, maybe with their whole group. The same plank coffee table and omnipresent bookshelves. But a strip of Arab embroidery enlivened the table, and lighthearted colors sprang from paintings, possibly by gifted friends. Family photographs were everywhere. My father's rooms had shunned color. He'd certainly owned nothing like the baroque silver menorah, poised on a silver dish, that turned his friends' small dining alcove into a sanctum.

Leaving me with her husband, Sara went to dress for dinner. Uri, wiry and sandy-haired, with a catlike chin, gave me a very sweet smile and attempted to converse. He'd never heard of the school where I taught, but had colleagues at the University of Michigan, geographically close. This consumed a few minutes, and persuaded me that I lacked the one kind of class for which the Gils cared. Close to tears, I chattered about my education at better-known universities, while Uri tried not to think about nuclear physics. Once or twice, his hand reached toward a stack of *Physics Today,* then wavered harmlessly away. He solved the problem by grabbing a Rubik's Cube and manipulating the plastic puzzle with a sound like cracked knuckles.

"Do you share your father's ideas?"

"Depends . . ."

"Phelps, for instance."

"I don't think Phelps is controlled by alien beings."

"Phelps drove us around Tel Aviv blindfolded, did Itzik tell you? Did he tell you how? No? There were strings in Phelps's blindfold, attached to his cuffs." Uri raised his Rubik's Cube to the light, carefully, as if it held a quark or two. "Your father and I used to have such talks, that lasted all night, about all kinds of wild ideas. . . ."

I asked if sometime, if he didn't mind, if he could tell me what he and my father had talked about.

"Perhaps."

He gave me my first security smile. I didn't know that yet; I took it personally, as a check to my presumption. And fell mute.

"I was telling Sharona about Phelps," Uri acquitted himself as Sara returned in a smart lavender suit. He looked at her and she looked at him. Her eyes darkened with warmth.

"Speaking of nonsense," she said, "I attended a forum this afternoon on the Gaia hypothesis. 'The planet is a living organism.' What *hantarish*."

"It might be arguable," said Uri, "if the planet reproduced, and made little planets."

"Does Sharona know about the plastic bags?" Sara cued. "I'm going to warm up the cutlets. Misha will come in a minute if he's prompt."

"If" was Uri's dark aside. "Your father and I," he continued, "had an idea to stop an Egyptian invasion. We drew up a proposal, and the brass sat on it, and sat, and in '73 they were sorry. But that means I can tell you. We were going to line our side of the Suez Canal with giant plastic bags stuffed with cordite. "

"An explosive!" I was happy to show off the homework I'd been doing in my landlord's library.

"A propellant. Though it can, under certain conditions, be used as an explosive." The Rubik's Cube crackled. I bit my lip. "If the Egyptians began crossing the canal, the cordite would be electrically ignited, and the gases of its combustion would puff up the bags, displacing tons of water. Thus creating a tidal wave, as it were. The enemy would be swept away, 'chariots and horsemen!' Our inspiration was Exodus."

"Wow."

God, I sounded dumb. I wasn't used to my father being gone—all over again. I missed him all over again.

"Itzik always thought about security. After he left for the States too," Uri added meaningfully. I caught a glimpse of Sara circulating in her kitchenette, speaking into a cell phone tucked between ear and shoulder.

"Matti, guess who's here in our salon? From the United States. You will never guess! The daughter of Itzik."

Something oniony sizzled.

Part Two: Of Visions and Illusions

Four

WHAT A SHOW! The bay window is black. In its fifty-five white square frames are fifty-five ruddy fires. I've lit a fire in the study hearth and look how many erupt against the night; true reflections, down to the smallest flame-fangs and the spark-burst whirling up the chimney. Tranquil in their geometric sequence, yet fey, as if my little fire were dancing on the dark facets of some implacable mind. A god of war to whom all are exactly alike.

I'm examining my little Passover photograph, whose folks are very much not alike. Yehuda, for example: he's the tallest, but what makes him stand out isn't height, it's focus. Even with a breeze mussing up his hair. His hands are clasped at groin level, feet planted in parade rest; he can stand there as long as he likes. Let the rest press and fidget and grin. His body says to the camera, "By all means, Time, my friend."

We were to meet on the top floor of the Sun Spire, the solar energy lab at the Weizmann Institute of Science. I took a footpath under the cavernous domes of the ficus trees, thinking about a slim, privately published memoir called *Hemmed Days,* which mentioned these aged trees. Fifty years ago they had given the author, Shai, a young physicist, a profound sense of mission. Off the footpath, out of the shade, stretched a gravel lane between rows of cypresses. It was noon: the white gravel flashed, the black cypresses glistened with powdery light. At the lane's end rose the steel needle of the Spire. I'd have dashed there if I'd had the strength. Instead I walked and walked the lane's solar gauntlet, green spots riddling my eyes, heat stewing my lungs, the handbag chafing my shoulder like a slow brand. I reached the Spire's entrance and stood panting in the empty, air-conditioned lobby, until my hairline dried. Then I found the elevator (also empty) and rode to the top. The doors slid back.

A vast loft. Filled with a cacophony of screeches, crashes, and thunks as the far wall split, floor to ceiling, in a narrow slit of light. A dark figure moved slowly across the loft and shook my hand, once. His face was sunk in his neck, and had the wildish, wise-ish look that the intelligent get from a cast in one eye. Both eyes took time registering their first impressions.

"You look like your father. Come, I'll show you the mirrors."

We walked to the slit in the far wall, our footfalls echoing on after we stopped. We looked down into a clearing of very bright dirt, where oblong mirrors on pillars, in concentric rows, stood tilted upward with a sacral stillness.

"They're being repaired," Yehuda said. "In operation, they focus the rays of the sun into a sheaf, a super-ray, that shoots through this slit where we're standing into the solar reactor. There the super-ray gets very hot." He rummaged in his trouser pocket. "This hot."

He held out a stone disk, its center melted to a stone drop.

"That ray—how many degrees is it?" I edged away, and Yehuda led me to a pair of metal chairs.

"In some experiments," he replied, sitting down slowly,

44

"nearly the heat of the sun's surface." His pauses had the weight of uttered thoughts. "But we cannot cheat nature. We cannot concentrate the sun's light until it is hotter than the sun from which it came. That is against the first principles."

"Oh! Are there really first principles?"

He turned up his cushioned palms, eloquently. I took out my notepad and he smiled.

"You've come to the right person. I was there at the precise moment your father arrived in Hemmed." He seemed to linger over a trove of stories. "One morning between ten and eleven, I was working in the Studio when the door opened. The Studio was our headquarters; it was a shed on top of an apartment building. I looked up and saw a tall kibbutznik, strangely happy, with a jute sack on his shoulder. He untied the sack and took out a Loretta. It looked handmade, like a toy."

"Yehuda, in the army dictionary, the term for that weapon is *Tolar.* Why was my father's called Loretta?"

Yehuda thought.

"The Italians had a Beretta, I suppose we could have had a Loretta. The point is, Genka was excited and liked your father's device. Genka was our director. As for your father—" An earnest silence. "I will have to use a word in English. Itzik was always— *exuberant.*"

"That's right, that's him!" I waited, but it was over. This was why I'd been invited to the Sun Spire: so that Yehuda could place one memory into the mind of the person proper to it. A believer in first principles, a keeper of the world's decorum, restoring it in this small way.

On the other hand, when Yehuda related his own Hemmed story, it was like a door opened onto the biggest picture I had seen of the '48 war. Yehuda had invented the Invisible Mine. The story of this weapon was Promethean: it was about the flame of scientific creativity turned into the fires of war, and how that transmutation broke the heart of Israel's first president, Chaim Weizmann,* founder and namesake of the Weizmann Institute. Despite my father's frustratingly brief role in the tale—a kibbutznik glimpsed once, strangely happy, in a doorway—I grew to know him more profoundly, as a man of his time. But I had to fill in the background, for Yehuda's story came alive only against it. Luckily, I had the library and the works of my distinguished historian landlord, to which I resorted frequently. For every blank page in my Hemmed notes, there were three or four volumes, bookmarked with bars of halvah, lying around the apartment. It's one way to expand your mind—wanting to know your father so much that you'll wade after his elusive presence through the chapters of A History of Zionism. To know the world that made him, who made you, with all its quaint and dreadful powers; its top-hatted statesmen and walrus-whiskered savants and generals in mufti, its treaties, philosophical treatises and soot-belching trains, and hosts of minuscule figures like your father and yourself, whom only love enlarges.

Chaim Weizmann, the center of this wide-flung tale, was enlarged by the love of the entire Jewish people. For good or ill.

For a while.

To tell the story of the Invisible Mine, which is Yehuda's, my father's, Hemmed's, Weizmann's, Ben-Gurion's, Israel's—I push

---

*Throughout this book, I have used the English letter *h* to correspond to the guttural Hebrew consonant usually rendered as *ch* or *kh*, and pronounced like the *ch* in the Scottish word "loch." This was done to avoid spelling the name of Hemmed as "Chemed," and having it pronounced like "cheese." But in the case of Chaim Weizmann, I did not presume to change the spelling preferred by *Merriam-Webster's Collegiate Dictionary* and the *American Heritage Dictionary of the English Language.*

slightly ajar the weighty coffered door of history. Here is the big picture into which my father's small kneeling figure, in the Passover photograph, fits.

My father was not a usual man, but the one really embarrassing trait I grew up with was his pretending to spit at the name of Britain.

"*Brittim—Tfooo!*" he would go, turning heads in the ticket line at the Boston Museum of Science. No amount of "Daddy, please . . ." would correct it. It's rather a long story.

From 1920 till 1948, Britain ruled Palestine under an international agreement known as the British Mandate. Weizmann enters our story in 1917, at the age of forty-three, just as Britain won Palestine from the defeated Turks. Weizmann was a chemist whose work on acetone had been vital to the war effort. He was also a Zionist leader, a future president of the World Zionist Organization. Cosmopolitan, charming, eloquent, with his diplomatic talent he moved the British foreign secretary, Lord Balfour, to draft the Balfour Declaration supporting a Jewish national home in Palestine. Such is the portrait of Dr. Weizmann in histories. My own sense of his character comes from a passage in his autobiography. He wrote to quash the notion that Palestine had been handed to the Jews in reward, as it were, for acetone.

I almost wish that it had been as simple as that, and that I had never known the heartbreaks, the drudgery and the uncertainties which preceded the Declaration. But history does not deal in Aladdin Lamps.

This sharp dictum is very hard to reconcile with Weizmann's loathing of Hemmed and their Invisible Mine.

The Invisible Mine was a product of Israel's three wars of independence, more or less, fought in overlap. Now, I'm going to show you a montage of World War II in Palestine. And there's something missing.

All over the country, on walls and kiosks, posters show Winston Churchill scowling above brave words in Hebrew that he seems to utter himself. From a community of 600,000, some 27,000 enlist. Two native army scouts named Moshe Dayan and Yitzhak Rabin go to fight the Vichy in Lebanon; Dayan loses an eye. Café Empire raises its black-and-gold sign, beer flows, the Aussies are our best mates, and tunes wafting from British open-air bands are picked up and hummed by kibbutzniks. Mussolini bombs Tel Aviv. Four thousand Jewish women parade in the A.T.S., chins up, in unflattering midiskirts, while 140 of our troops are lost with a bombed British transport ship. Rommel crosses into Egypt—suddenly the British and our illegal defense movement, the Hagana, are jointly training commando units called the Palmah. If Rommel wins, the Palmah will cover a British retreat and hold the ground. Our 2nd Infantry is posted to Libya, our 1st is smoking Egyptian cigarettes. The Jewish Brigade is formed, hurrah! Five thousand troops sail to southern Italy and see action. A sassy song: *Hoi, Emperor Titus! If you only saw, beside the Arch you built, the soldiers from the Land of Israel!* A joke: when the wounded on the battlefield are crying "Water! Water!" the brigade men cry, "Seltzer! Seltzer!" Hanna Senesh, a parachutist twenty-two years old, is jailed, tortured, and executed in Budapest, her birthplace, leaving behind several poems that become long-lived songs: *Blessed is the match that burned, to light the heart's secret places. . . .*
Genka Ratner, Hemmed's future director and the only quali-

fied inventor in the group, joins the British navy and invents the limpet mine.

What's wrong with this picture?

While you're thinking, I'll tell you how two other readers answered the question. My husband, a Methodist from Atlanta, said, "I don't know. It figures the Jews would fight with the British, what with Hitler and all."

An Israeli friend of mine said what Israelis always say. She said, "Are you crazy?!" Then she began arguing: "A book? You're making a skewer, for the leftists to skewer you from the left, and the rightists to skewer you from the right, and what you'll be, sweetie, is shishlik." Then she demanded: "I want you to put in about how the British sent my dad to Eritrea because they searched our house and found rifles in the closet." When I promised to, she answered my question. "What's missing? How we threw the Brits out!"

Here it is: Israeli independence war #1. Winston Churchill doesn't want it. In the parliamentary debate on the White Paper of 1939, he orates from the back bench: *I cannot understand why this course has been taken.... Is our condition so parlous and our state so poor that we must, in our weakness, make this sacrifice of our declared purpose? ... Never was the need for fidelity and firmness more urgent than now.... By committing ourselves to this lamentable act of default, we will cast our country, and all it stands for, one more step downwards in its fortunes.* He quotes Chamberlain's speech on the Jewish national home, twenty years earlier: *It is twenty years now that my Rt Honourable friend ... used these stirring words: "A great responsibility will rest on the Zionists, who before long will be proceeding with joy in their hearts to the ancient seat of their people. Theirs will be the task of building up a new prosperity and a new civilisation in old Palestine, so long neglected and misruled." Well,* Churchill concludes, *they have answered the call. They have followed his hopes. How can we find it in our heart to strike them this mortal blow?*

The blow falls. The White Paper of '39 cuts Jewish immigration to Palestine to a trickle, for the coming five years—then to a stop. Despite the noose tightening on the neck of European Jewry, immigration is already restricted: in 1936, out of the 22,000 immigration certificates requested, only 2,500 were granted. Now the gates of Palestine are closing. Illegal refugee ships appear in rising numbers on the sea, and the British Coast Guard shoots at them. Turns them back onto the high seas, often to sink—these are run-down hulks, overloaded with desperate people. The *Struma,* turned back and torpedoed, goes down with all 769 passengers. The *Mafkura,* sunk by the Germans, with some 400. The *Sakaria* is caught by the British, and its 2,300 passengers are put in camps constructed for the sole purpose of keeping refugee Jews out of Palestine. The camps are stark and pestilent: on the island of Mauritius, in the Indian Ocean, many refugees die of tropical diseases. Britain pursues this policy for reasons on which I am no expert. My historian landlord suggests Realpolitik—appeasing Arab countries, the many trumping the few. Or responding to Palestinian Arabs, who view Jewish immigrants as a demographic threat to their own nationalist movement, and cannot see why their own dream of an Arab Palestine should be sacrificed because of the crimes of European fascists. In their shoes, I might feel the same way. In the Mandatory government's shoes—but what a farce, to place myself, rhetorically, in the position of officialdom! The authorities claim that it is dangerous to shelter Jewish refugees: they could be Nazi spies.

.   .   .

My father is nineteen years old. He lives in a tent on a sandy bluff, overlooking the sea, among the tents of his first kibbutz. At night, with everyone else, he sneaks out on word-of-mouth missions to land illegal ships, or ferry the people ashore in rowboats, or help them swim the distance, then hide them on the kibbutz. He does not have a monument in his head labeled HOLOCAUST and engraved with names. He has ugly rumors, and the first— surely questionable?—newspaper headlines in 1942. I think he must tell himself certain things on the way to a mission; and certain other things as he returns, in fierce, ignored, interior whispers. I think he must look for his family in every pair of reaching arms.

Weizmann speaks to the Zionist Congress in 1939.
   We have not failed, he says, *we believed in Britain.*
   But Britain no longer believes in the Balfour Declaration. Leftist delegates to the congress argue that they should never have trusted an imperialist power. Rabbi Berlin advises trust in God. Weizmann ends the conference: he speaks of endurance, of thick gloom and a future light, the light of peace. Many delegates reach to embrace him. He has their hearts. But another voice is walking into the minds of the delegates, slamming windows, opening doors, switching the cross breezes: . . . *the homeland is not like the Diaspora. Here we do not stand by helplessly. This is the one place in the world where the Jew will stand and fight.* This is Ben-Gurion. He is interested in results. *He will fight and he will succeed.*

If the casus belli of war #1 is the White Paper, the battleground is the law against Jewish possession of weapons. The Hagana, the Jewish defense movement, is illegal; right through the world war, British police conduct searches, confiscate weapons, and impose severe penalties. The doublethink is tricky, requiring a juggling act performed with riven loyalties. It means fighting for the sake of a government that doesn't care to protect you from the enemy, but on the contrary, acts to increase your peril abroad and vulnerability at home. The same period that sees your 1st and 2nd infantries dispatched to Libya and Egypt also sees your recruitment office (of all places) in Tel Aviv ransacked for illegal weapons. And a kibbutz searched so brutally that one member dies and dozens are wounded. Your newspapers, which condemn the police action, are gagged. As your first paratroopers drop into occupied Europe, you are sentenced to seven years' jail for the possession of one bullet in your pocket, and you happen to be Dr. Weizmann's bodyguard. Weizmann is widely regarded as the ambassador of the Jews. Quite. And your Palmah—the same Palmah created to cover a British retreat and hold the ground, waging guerrilla war for as long as it survives—is promptly stripped of its arms after El Alamein.

The Palmah, thinking it over, in March of 1943 raids a British arsenal, carrying off twenty-two machine guns and a couple hundred rifles. Which (they say) belong to us. They do more than smuggle refugees. They blow up British coastal patrol launches, maritime tracking stations, radar installations, police stations, and railroads. The Jewish rightist fringe does worse, with hostage abduction, bank robbery, hotel bombing, and political assassination (to the fury of the Palmah, stuck with damage control; rightists are on my father's spit list). The Mandatory government begins to run amok. It is using tanks and thousands of soldiers, German shepherd dogs, and metal detectors to search this village and that kibbutz, tear up tents, throw mangled mattresses on the ground, rip out floors and walls, drag angry women by the hair, lock protesters in cages, shoot people dead, cart off thousands more to jail to be tortured and beaten. Tel Aviv zigzags from raging demonstrations to rigid curfews and mass interrogations. With the end of World War II, and into 1946, independence war #1 is in full swing.

It's hot, where my mind is. My body is enjoying the coolness of the Spielberg film archives in the basement of the Hebrew University. Blinkered by the extended sides of a view screen, running a newsreel, I am at the Weizmann Institute, in '49, and man, it's hot here. Today is the inauguration ceremony, in a flickering pavilion tent with sun-beaten sides. On a platform, in his shirt-sleeves, Ben-Gurion rasps into a microphone, his pugnacious chin and underbite on full display. Dr. Weizmann wears a dark suit, his expression hidden by dark glasses. His vision is deteriorating. He stands like he knows he's going to lose. Like a post. The Sieff Research Institute is being renamed for him. This honor isn't power.

I want to tug his hot black sleeve, and say—something comforting.

It's 1946. One hundred thousand stateless people, mainly survivors of concentration camps, languish in Displaced Person camps run by the British government to keep them out of Palestine. Truman presses on their behalf.—A little story about the DPs that my mother heard from my father, who had pondered it for ten years. His kibbutz smuggled in some DP children. He carried one sick little boy to the clinic. The rest ran after him and crowded in the clinic door, hitting one another, muttering in

German, "Shoes, shoes." Evidently, they were not used to seeing a child enter a clinic and come out again alive. They were jockeying for the dead boy's shoes.

Ernest Bevin, high commissioner of Palestine, at a Labour Party Conference, explains why the Americans are pressing for the DP Jews to be admitted into Palestine. They "did not want too many of them in New York."

The Palmah, thinking things through, goes out one night and blows up ten bridges connecting Palestine with neighboring countries, losing fourteen lives in the process.

At the Zionist Congress in Basel, when Dr. Weizmann speaks of endurance, restraint, the political temperature freezes his breath. Passionately he condemns the path of violence: *If you think of bringing the redemption nearer by un-Jewish methods, if you lose faith in hard work and better days, then you commit idolatry and endanger what we have built.* He is not reelected president.

Here it is: independence war #2. The United Nations votes, on November 29, 1947, to partition Palestine into two states: one Jewish, one Arab. In the following days, militant elements of the Palestinian Arab community express rejection of the Partition vote in several attacks, which escalate into the civil war between Palestine's Jews and Arabs.

You might wonder about Hemmed's attitude toward the Palestinian Arabs. Those scientists whom I met supported the peace process as Itzhak Rabin was then pursuing it. My own leanings were toward the leftist position that advocated a Palestinian state for reasons of security and national conscience; indeed, my research was made easier by knowing that Hemmed had operated chiefly against the Egyptian, Syrian, and other national armies, rather than local Arabs. And as I talked with friends about the violence in the West Bank and Gaza, that summer, I would catch myself praying that Hemmed's struggle for a Jewish state, which took the uttermost they had to give, would not be wasted, would not be for this.

But understand: that tight-lipped group was talking to me only because of their affection for my father. I was not going to jeopardize their ease with me for the sake of political debate. When Misha insisted that not one Arab had been induced to flee in '48, I raised a mental eyebrow. When Uri said that, after all, the Arabs had chosen war in '48, and "one does not go to war with an insurance policy," I nodded. Privately reserving the thought that in the factionalized Arab community of '48, those who went to war did not always represent those who paid the war's price. I knew better than to credit the myths of blameless heroism, but I wanted to capture the texture of this group's experience. Recent historians convincingly show that the notion that Israel fought purely defensively in '48 is a myth. Ben-Gurion, they show, used the war to push for territorial gains— but from Hemmed's perspective, the stakes had been survival. "I didn't like drawing weapons," said a famous architect who had been a draftsman in the Studio. "I am not a party to destruction." But when asked if he felt guilty: "No. We were so small, it was clear." I remember hearing my father say, "I feel sorry for the Vietnamese, they are little, like we were."

Three months remain till the birth of the Invisible Mine. The law against possession of weapons is enforced, at police road-blocks, against every Jewish passenger bus or farm truck climbing to the city of Jerusalem. Arab attackers effectively turn that wooded hairpin mountain road into a fatal obstacle course, preventing the transportation of food, people, and supplies. War #1 and war #2 merge: between them, the Jews of Jerusalem are besieged.

Yehuda studies physics at the Hebrew University, where his story, as I heard it, begins. The campus connects to West Jerusalem via a road that passes Arab neighborhoods, making every bus a target. When war #2 erupts, school closes, and the students go a-soldiering. Professors stay home and boil edible weeds plucked in the park.

But in the deserted university, a small group of students camps out in the Physics Building. They sleep in the labs, on army cots; Yehuda beds down beside the optic spectrometer. They open locked reagent cabinets and realize, suddenly, that no one is going to suspend you for unauthorized use of the sulfuric acid! Or anything else! I don't know what they're eating, but they have a great time. They are the core of Hemmed-Jerusalem. Their director, brilliant, idealistic Aharon Katzir, tries to organize them on the kibbutz model of rotating tasks. True scientists, they shudder and form specialized research groups, mechanical and electrical. Their commander, eminent Professor Racah, sets up headquarters in the vacant switchboard office, where he is found at all hours deftly plying his slide rule. He is in the throes of inventing Hemmed's most beloved device, culled from an Italian journal. It was loved for its intrinsic beauty, as it is also the only Hemmed device to enter English-language histories for failing in a major battle. The Hollow Charge, *la carica cava,* Racah lectures, with his trilled Italian *r*'s, is an explosive charge molded around a cavity, which causes the blast to pierce armor. A beautiful idea! *La carica cava* triumphs over brute firepower, the advantage of our enemies, by means of—*cosa? Niente!* A hollow cavity!

Empty space!—and the laws of physics. *Magnífico!* The troops are on fire to make the thing. Off they go to the lab. A block of wood is hollowed into a bowl shape, TNT is cooked in a saucepan. It catches fire. It is dropped on the floor. The lab hastily empties of personnel. A quaking student throws a towel over the flambé.

Hemmed lives to fight another day.

Yehuda spends hours in the blacked-out National Library, reading up on machine guns and mortars, propellants and explosives. The Invisible Mine does not yet occur to him. Until one morning he is thrown from his cot onto the floor. Propping himself on his palms, he looks down at his wristwatch and shattered teacup, which he left overnight on the windowsill. It is 6:30 a.m. The continuous roar is so overwhelming that he cannot hear around its edges. He takes a few seconds to connect it with the broken china.

Shortly before the British leave Palestine, the colonial high commissioner holds a reception at Government House, on Government Hill. A freelance journalist attends. She notes the dignified Sudanese servants in dark blue robes with red sashes, circulating with trays of cocktails, salted almonds, pretzels, pickled gherkins, stuffed olives, and caviar canapés. She samples each hors d'oeuvre, and describes them in her personal diary. Her name is Malka; in her thirties, single, snobbish, pretty. During the siege, she humorously rates the barter value of her shrinking figure, and records signs of malnutrition. "Round little wounds, that neither hurt nor bleed, open on my face, lips, and neck. . . ." Dizzy spells. What she needs is vitamins. She thinks about clothes. Foreign fashion magazines still crop up here and there, and Malka wants the New Look. She stands on a chair in front of a small mirror, rather than risk the full-length mirror in the room upstairs, which is more exposed to shelling. A hem could be let down, lengthened with a piqué towel! Her flat black sandals are stylish again, the ones she bought in Cairo. There she was a nattily turned-out WAAF with the Royal Air Force.

At 6:30 a.m. on a February morning, she is sprinting toward Ben Yehuda Street. She turns the corner and sees that the people who have not been crushed by fallen buildings are wandering around in pajamas, some clutching their clothes but unable to dress in the midst of howling crowds. Families in pajamas, on mounds of rubble, are screaming names into holes they dig with their fingers. Stretchers push past, slathered with bright red, and she turns away to find, at her side, a speechless little boy. Someone has pulled a Persian lamb jacket out of a shop window and draped it over him. Malka picks a glass splinter out of his brown cheek, another, a third, with impatience a fourth and a fifth; then feels it's useless. Walks off. Grabs elbows, demanding information. Some fifteen minutes ago, various people tell her, an armored car followed by a truck pulled into Ben Yehuda Street. The car pulled up to the sidewalk. Some tall blond Englishmen got out. Without stopping a flow of loud profanities, in front of the stupefied bystanders, they attached a fuse to the loaded truck, dove back into their car, and raced off. Then the street blew up.

A woman in curlers rushes up yelling that the sandwiches handed out by the rescue workers aren't sandwiches at all, they're an insult, look at them, look. An obese butcher in an apron leans against a wall, burping, vomiting, and wiping tears with his broad fists. Farther up the street, Malka passes an armored car, its gun trained on the crowd, in which a young subaltern with the shoulder flashes of the Royal Engineers complains to his sergeant that the rescue workers are doing it all wrong. He looks ready to cry. Malka stops, delivers a verbal slap, goes on walking. Finding herself in a stationery shop, she folds a megaphone out of construction paper, tapes it, runs outside, and hands it to a hoarse rescue worker. The blast has left a crater, she observes. Very tall flames nod from the ruins. Eden Hotel is a square of shattered glass. She puts her face in her hands, briefly; touches grit.

Café Empire tosses its sign into the street, and puts up a new sign, Café Ben Yehuda. Hemmed also reaches a decision. In Yehuda's words: "We had to stop British cars. We couldn't set up roadblocks and man them, it was illegal for Jews to man roadblocks. A land mine was the best alternative. But it was illegal to be caught with weapons, so the best mine would be invisible." I asked Yehuda what made the mine invisible. He gave a slow, considered, security smile.

"It was designed," he replied, "not to be seen." The mine is still in use today. Or so I'm told.

In March, Yehuda receives an army order to join the Studio. He leaves Jerusalem in a slow convoy, holding his breath intermittently; climbs out in Tel Aviv; and I'd like to think that he makes a beeline for the nearest restaurant. Reporting, with smothered burps, to the Studio, he meets Genka Ratner with his knife-pleated uniform and his black-light smile. Genka wants to develop the Invisible Mine posthaste and assigns Yehuda a teenage assistant.

"Today he's a famous architect. Then he was just a *tsootsik,*" says Yehuda.

This stripling has the effrontery to suggest that the mine ought to be round, with parts fitting inside the lid.

"He was right. It could be manufactured more easily if it was round; it could be spun. But I was irritated. When a thing works, you don't make a Platonic symposium about it."

The state of Israel is declared by Prime Minister Ben-Gurion, in a speech broadcast over the Studio's miniature radio, a novelty the size of a jam jar; inside it, BG's buzz-saw timbre is diminished to that of a bee. (Epochs rise and fall on the fulcrum of traceless minutiae: let all such fateful speeches be heard thus, in miniature!) On May 14, 1948, the United States recognizes Israel, followed by the Soviet Union and other countries. Britain stays officially ignorant of Israel's existence. Egypt's largest newspaper avers that the problem of recognizing Israel will be solved when Israel ceases to exist. Darkness fills the eyes of Commander No One, Hemmed's exec: by his reckoning, Israel's supply of rifles, machine guns, and light and medium mortars cannot possibly repulse invading Arab armies equipped with tanks and fighter planes and artillery.

The Invisible Mine is now a weapon of war #3.

Hemmed expands out of Tel Aviv into the town of Rehovot, and the Sieff (soon to be Weizmann) Institute. When Genka asks Yehuda what he needs to make Invisible Mines at the Institute, Yehuda says, "Four girls and a room."

He is assigned four ponytailed troops and a second-floor office in the Zisskind Building. This is all so long ago that WEIZAC, cousin of the ur-computer ENIAC, has not yet been built, and does not fill a showcase in the lobby with all its dust-furred rows of cathode tubes. But the floor has been laid and polished, and the wooden stairway curves toward the small white-hot skylights of the second floor. At the foot of the stairs, Albert Einstein's bust presides on a pedestal. Speaking of art and science, our own Dr. Einstein takes pride in a cartoon poster he has drawn himself, showing four baby girls and a passable likeness of Yehuda, frazzled, boiling diapers. It is captioned: "Yehuda's Kindergarten." Of course Yehuda can't resist posting this masterpiece on the pebbled-glass door of the Invisible Mine workshop.

Where Dr. Weizmann can't help seeing it.

On the last leg of the march begun thirty years ago with his efforts to win the Balfour Declaration, Weizmann lobbies Harry Truman to recognize the state of Israel. This achieved, he receives the news, in a telegram, that he is Israel's first president. Yet Weizmann returns to Israel a man shorn of political relevance. The presidency is a symbolic post. Prime Minister Ben-Gurion is the man at the helm. I am trying to understand the meaning, to this old, wise councillor, of his Institute. Of all the visions he has nursed into realities, this is the one remaining in his hands: his Institute, a pure vision uncontaminated by his bitter rival, Ben-Gurion, who has not even let him sign the Declaration of Independence. Weizmann lives on campus, ailing and near-blind, in the atmosphere of his first, scientific calling; among dreams of fizzing test tubes transformed into hyper-grapefruit and hybrid wheat, into meals in India and vaccines in China, into a final, global appreciation of what the Jewish mind stands for.

But grapefruit go missing from the agricultural school's grounds. On the other hand, soldiers pop up everywhere. The Institute's rural hush, heightened by a birdcall, the sawing of a cricket, and the monastic hum of intellection, is broken by thuds. The usually placid chef of the faculty club bursts into his office, distraught as only a chef can be; she flaps her apron, and floury clouds rise. Plaster, she cries, plaster raining from my kitchen ceiling! The rotten boys with their explosions, they do it on purpose! Mister President Doctor Weizmann! *Do* something.

He forbids Hemmed to eat in the faculty club or sleep on campus. His authority extends that far.

In July, Weizmann writes to his close friend and aide Meyer Weisgal: ". . . the entire scientific basis of the Institute has been changed. This was done without the least consultation. . . . There is no justification for the whole scientific enterprise . . . being exchanged for something that isn't science at all, but the manufacture of explosives. . . . The loathing I possess for all these military operations is so deep, so rooted in me from childhood, that, plainly and simply, it grieves me to see the Institute, to which I've dedicated so much energy and so many effortful hours, become desecrated in a way which I cannot explain and with which I can make no peace."

Yehuda is supervising the mine workshop when a knock comes at the door. He opens and wants to close it. Weizmann with his goatee and necktie and black suit looks in, beside nervous Ernst Bergmann, the Institute director, who gets on Yehuda's nerves— the genius type who lights a cigarette, forgets, lights another one, smokes them both. Now he hasn't steered Weizmann clear of the office and its foolish poster, which is bound to make the president ask, "What is going on here?"

"Sir," Yehuda murmurs. His four girls' hands rise to their lips. Weizmann shuffles forward, forcing Yehuda to back up. Removing his dark glasses, squinting with naked eyes, surprisingly Slav, the president asks, "What is being produced here?"

Yehuda looks at Bergmann. Yehuda looks down at the elderly gentleman, the Jewish nation's ambassador to the world.

"Mines." He sees no point in calling them invisible. The president begins trembling; he speaks in a clear, shaking voice, as if to a formal assembly.

"This is a scientific institute, not a munitions factory."

He can never remember what transpires until Weizmann is led away, amid Bergmann's complicated apologies, and Yehuda shuts the door and leans against it. The girls' hands are still lifted, but not covering smiles. One of them, hushed and angry, breaks the silence.

"But the Egyptians are bombing Rehovot."

How can he explain to these youngsters what a research scientist feels about the perversion of his sacred flame, his truth, into a destroying force?

"He knows," Yehuda tells her. "He knows that."

When Yehuda finished his story, I asked for a tour of the solar reactor. He willingly led me to a table, which was all the space it needed. One component was a ceramic cylinder. Over it hung, upside down, a bowl-shaped mirror. From these two parts ran a tangle of scorched and blackened gooseneck pipes, into sundry containers.

"The idea is to create a miniature sun. Remember the super-ray?"

Indeed; and I saw that Yehuda would always smile at how I'd backed fearfully away from the slit in the wall. When the super-ray shot into the lab, he explained, it was already focused to a burning hot point. Why not refocus that point—which was an image, but also an object in its own right—to achieve a double concentration?

"The image is also an object," I repeated, struck. Yehuda's

split gaze turned doubly whimsical, and we smiled at each other through the tangled pipes.

"The image is also an object," he confirmed. "The burning point can melt stone. It is a thing in its own right. It can also be refocused, like any image made of light. We refocus it here, in this bowl." He touched his inverted mirror with a knuckle; it was no bigger than an ordinary bowl for mixing eggs. Yet a miniature sun would blaze and shine within, like ours in the dome of heaven. "That at least is the vision."

"Do you mind my asking . . ."

Perhaps the speaking quality of Yehuda's silence came from his ability to wait, head sunken in his neck, like a sea turtle or some sempiternal creature.

"I was wondering, how do you know when a vision is real and when it's just an illusion to begin with?"

"You assess the facts and probabilities. Then you work, and you wait."

"Time is the final test, then?"

"Time." Slowly, with a big, crooked hand, he indicated his cup, his mirror, and his pipes. "The only way to work on such a project with a quiet heart is to accept that you will not finish it." He gave me the kindliest look. "That is for future generations."

Five

ON JULY 6, 1995, I called Kibbutz Shoval and spoke to the archivist. I said I was doing family research, my father had been a comrade in the forties. I said nothing about his career in weapons development; the Israeli scientists who assured me that my father had come to their secret lab straight off the farm also believed that the kibbutz wasn't privy to classified matters. Though, as everyone knew, few secrets were safe from old kibbutzniks.

"Your father's name?" asked the archivist, and I told her. There was a long pause; I thought she'd gone away to open some dreary file cabinet, where my father's name might linger in faded pencil on an old work-rotation schedule. Then she said, "I'm going to faint. The daughter of Itzik Bentov."

"You knew him?"

"Itzik Bentov's daughter. Everyone will faint. Ziva especially will faint."

"Who is Ziva?"

"You don't know? Ahh. Interesting. Ziva was your father's wife."

A glass of blue thistle tea. Beside it, a pastry roll on its tin plate, sweating beads of honey. The porch swing's awning cast an orange shade on the warm cotton seat. Ziva, being short, had to rock the swing with her pointed toe. A burst of black veins threaded her trim calf. She wore white shorts and a halter top whose cream stripes diverged over her bosom, and sitting erect, fingered a gold locket that rested in the dip between her neck tendons. Her posture, the straightforward yet quaint way she spoke, her vibrant voice free of all grudges but humming, somehow, with deliberate reticence; all these qualities I admired.

At first, I had seen only a messy little woman with slit, arched eyes in a pear face. After a moment of mutual staring, I saw that the sun had leached Ziva's face, leaving a waxen skin inlaid with tan speckles, which gave her an untidy air. About the same time I revised my first impression, Ziva, facing me in the back seat of the kibbutz station wagon that had fetched me from the bus stop, reached out and tapped me gently on the wrist.

"Two drops of water," she said. "You and Itzik."

Her home might have been any Israeli farmhouse; a red-roofed bungalow with a satellite dish, though by kibbutz standards, the five rooms were luxurious. Photographs of her numerous grandchildren were distributed over the bookshelves. She dabbed one with a fingertip.

"Negev dust. Wipe all you like, nothing helps. Shall we drink indoors or on the patio?"

For once, I wasn't shy. I was openly snooping around her kitchen, where Bedouin necklaces of pot-metal charms dangled from a nail by the sink. Ziva, like me, adorned the spot her eyes rested on while she washed dishes. I said that I'd rather sit outside. She said, "Ah! I too."

· · ·

As we sat drinking thistle tea in the orange shade of the porch swing, a three-walled hedge cut off our view. The hedge made our talk intimate, though I knew that just outside these jade-leafed walls another world extended: the yellow sand swells traveling to meet the sky, which was wider and farther flung than the desert slumbering beneath it.

"It's good to see the Negev sky again."

"Your father also loved it." A corner of her lip curled into one wrinkled cheek. "When we first settled here, in the sands, you could always see rubbish in the sky, bits of paper or string twirling around, or the sand devils. They went spinning up to unbelievable heights. That was the only difference. The land has changed, but it's the same sky.—You are a positive replica, *my dirr*" (she said archly in English). "I can't get over the sight of you! Itzik in a feminine edition."

"Would you mind telling me when you met my father?"

"Why else have you come? I knew him from Slovakia."

"I had no idea," I said, after some moments. "No idea. Ziva, all my life, I've thought that anyone who knew him from that time was dead."

"We met on the last train, and the last boat," she contradicted, her eyes sparking. "All the kibbutz was on that train. We did not know it was the last. How could we? Such a thought never entered our heads."

Chance it was, random chance, that saved my father, his wife, and the other founders of Kibbutz Shoval from the death camps. On the eve of war, in Nazi-occupied Prague, the British consulate was on the verge of closing when a Jewish Agency clerk went to an appointment there. While he waited, the clerk saw a stack of blank immigration certificates, unguarded, on a desk. These, he stole. Ziva met him decades later and heard the story.

"One doesn't envy him. He had to choose, among tens of thousands of cases, who would have life and who, death. And he saved us, because in the end he chose some hundred youngsters in the Zionist movement. That is one more reason, Sharona, why I believe that life is a game of chance."

Chance, destiny; they can be the same word, in Hebrew. A game of destiny. Some hundred Slovakian teenagers boarded a train for Trieste, in March of 1940.

"We kept slipping off that train with Nazi police at every station. Great fun. Our camp counselors were frantic. In Trieste we were strictly ordered not to set foot off our ship, the *Marco Polo*."

"You've got to be kidding."

"*Nu*, we were kids," Ziva cackled, "on our first trip to a real city. Itzik and I sat in the cinema all afternoon. The manager tried to give us tickets for the evening show; he wanted to give us free tickets. Can you guess why? Because of Itzik's laugh. A laugh would roll out of his mouth and people couldn't help themselves—you remember, he always sounded so surprised? People started laughing in sympathy, and it would spread, and soon he had only to begin a little peal and the theater shook. He made the movie much funnier than it really was. We didn't stay, we had that much sense, and on the way back to our ship we went up an alley. And in that alley were four fascist policemen in black uniforms, leaning on a wall, watching us. I wanted to slip by quietly. But Itzik ran right up to them, waving his arms and crying '*Cantare! Cantare!*'"

"Why?!"

"That's what he thought Italians did."

"And—"

"They looked at us. They surely saw, from our clothes, that we were country kids. Who knows what else they saw or knew. What they did: all four doffed their plumed helmets, held them to their breasts, and sang for us. 'O Sole Mio,' or something like that."

Ziva served me a chunk of the jam roll and asked, cocking her head, "Did Itzik convey to you anything of your Slovakian heritage?"

"Once he sang me a song about slivovitz."

"Slivovitz! . . . This pastry is called makosh. With chocolate filling, it would be called cacaosh."

"A song of three brandies," I recalled, licking my fork, feeling pampered. "'*Slivovica, borovica, terkovica, hey.*' I'd like some more makosh." Ziva cut me a generous chunk and went inside; she returned holding papers, which she laid out on the table's check-ered oilcloth. Her palms were solid callus, like my grand-mother's. The papers were xeroxed photographs; evidently, she was unwilling to part with the originals. She introduced the first picture with strained lightness, "The happy couple."

"Oh, how sweet!"

They stood side by side, before a row of shacks. The young wife's curved, mirthful eyes were alight like filaments. She wore a blouse with puffed sleeves, not a work shirt. My father was very thin, and his smile somewhat caustic. He already had a widow's peak—my widow's peak. I wanted to learn what had broken them up. I had to know, but was unable to muster words that might cause my new friend pain.

"We were a good-looking couple. Everyone noticed," Ziva said. "I was—not beautiful, but pleasing, and Itzik had that figure, that Greek shape! Between his shoulders and hips, that wedge!" She struck down graphically with her hands. "Yet I always felt that despite his sensational body, Itzik was not of this earth? That was my intuition. He was a hoverer, Itzik, he wished to soar. His body's lines were soaring, the way his brows swept up, the way his arm floated up from his shoulder, you remember? How he used to move like this?" And she imitated my father's gesture to the life.

"I had forgotten," I confessed. She leaned one elbow on the table, and gave me her lopsided smile.

"To tell the truth, he reminded me of the myth of Icarus. He wished to invent wings for himself and fly away. And he did, but the end was not good."

"No," I agreed.

"It's good you came," Ziva said, in a sort of crackle that struck me as very forties—like a radio bulletin. "You are the gift he has left us."

We heard the trudge of Moshe's boots and his bellow of *shalom*. He pulled up a chair and lowered his big frame, the sort of man who lacks all fuss but puts you, somehow, on your mettle. I learned later that his hobby was astronomy. He grinned at me, with gappy teeth, and served himself pastry.

"And what is the name of this dessert?" Ziva asked, suddenly the didactic mama.

"Makosh," I said.

"And with chocolate?" I shook my head.

"She doesn't remember because you didn't give her any! Wife! Why no cacaosh?"

"On her next visit we'll have cacaosh too. And that visit will be soon." Her tone, his attention, revealed that this visit had not been simple; and its success made me brave.

"I have to ask—did anyone with my name, on this kibbutz, ever commit suicide?" The elderly couple stared at each other.

"All the things Itzik didn't mention," Ziva declared, "and that story he chose to tell? That is extremely strange."

"Look, when I was about twelve he said to me, out of the blue, 'I named you for a girl on my kibbutz. She was called Sharona and she committed suicide.'"

Ziva clucked.

"What a thing to tell a child. Stupid Itzik!"

Moshe, who when he put anything down planted it, planted his hands on his thighs.

"You've heard of the Greenberg family," he began. "Not poor! Greenberg's daughter was taking some Americans on a tour of the Negev in . . . in . . ."

"After the war," Ziva murmured.

"Aha. So she drove up to the gate of the kibbutz in her big black *limoozeena,* and she drove away with Zalman, who was Sharona's husband. He didn't come back—Zalman didn't—so Sharona took a pistol and shot herself through the heart. In the haymow. They say that the husband of Greenberg's daughter also tried to shoot his brains out, but only got as far as his arm."

He finished with the ghost of a snigger.

That was the truth behind my depressing secret. I could have kissed it—for being real, for not coming from my father's eerie inner world, some suicidal nightmare he'd called by my name. No, I was named for a tale of early Israel! A vintage kibbutz drama, complete with capitalist harlot. Sharona's story was sad, but, frankly, I wasn't. I felt giddy with relief and a slightly hilarious happiness, as though I'd suddenly come into a fortune.

"Perhaps Itzik wanted . . . to make up for Sharona's loss, to us.

From afar," Ziva said delicately. "I am sure there was no . . . we would have known."

Moshe grunted; it settled the question.

"After lunch," Ziva suggested, "and after you talk with Zvi, our mechanic, we can take a walk to the graveyard, if you'd like to visit Sharona's grave. Unless you prefer to nap during siesta?"

"I adore afternoon walks."

"I too."

We walked to the dining hall on a pink cement footpath that rose through tropical verdure, toward the top of the kibbutz's residential hill. We skirted agave cactus like land octopuses, and crossed zones of head-swimming fragrance. Ziva told me that Shoval hosted conferences on desert landscaping. Over our heads bloomed the flamboyants, like red chilis on green trays. I stood staring at a blue puddle. There is no such thing as a water puddle in the Negev noon. It was made of fallen jacaranda petals.

The dining hall stood in a grassy clearing. We entered through a circular mail room, where eddies of people were chatting, into a sunny hall with vaulted windows. I got a cafeteria tray and stood in line; it was slow going, as the cooks would pause to chase, with a spoon, a particular boiled onion or chicken part that someone wanted. Ziva greeted everybody. I heaped my plate with cherry tomatoes. At a corner table, a man sat scrubbing his face with the heels of his hands, mobbed by people wanting to change their assignments on the work roster.

"And I, aren't I a human being?" moaned the roster planner.

I was smiling nonstop; joining my hosts at a table by the window, I wondered about this powerful euphoria. Could it be dangerous? If this was "up," what would "down" be? I thanked

Moshe for the olive oil cruet. Maybe I'd already been down. Maybe it had lasted thirty-nine years.

"There's Zvi," Ziva cried, waving. "Zvi! Over here!" The mechanic approached with a broken gait, a coffee cup in his crabbed hand. He wore indigo denim overalls, and smelled of motor oil and iron filings. "Remember the smell," Ziva said. "This was Itzik's smell!" Zvi had a droopy face with sad brown eyes, and a tired whine. He had to delay our appointment by ten minutes, to pick up his baby.

"I call her my baby though she is thirty-five years old. Down syndrome." Belatedly I stopped grinning. Zvi described the route to his apartment. "I miss your father very much," he said.

"I miss your father. We had such good laughs together," he repeated. His room was spotless, decorated with amateur photography, blurred close-ups of ferns. He showed me his daughter's room, which the kibbutz had built to accommodate her visits. I praised it, disparaged American health care, praised his photographs, but did not succeed in raising his spirits till I unfolded my notepad and asked him about the device my father had built called the Loretta.

"It was fun," he said, lighting up, illustrating the story with soft, appealing sweeps of his crabbed hands.

In the forties, Shoval's machine shop had had a partition dividing it from a storage closet. One day, Itzik and Zvi were testing the Loretta's homemade shell. Clamped in a vise, the shell was connected to a battery under the table, by a skein of steel wool that Itzik and Zvi had patiently unraveled from cleaning pads. The young men crawled under the table and made contact with the battery. The steel wool cable heated up, turned red, and the shell's rear end, complete with fins, blew off, broke through the partition, and hit the head of a kitchen worker who'd come into the storage closet to look for a bread pan. Unfortunately there was a slight concussion. She recovered, but the kibbutz council forbade weapons-testing indoors. Itzik's "electric gun" (as the comrades called the Loretta) would have been outlawed altogether but for the fact that Shoval's entire arsenal, to combat the Egyptian army, consisted of five rusty Italian carbines.

"You only had to look at them to know they wouldn't shoot," Zvi recalled. "If you tried, the recoil broke your shoulder."

Banned indoors, the two mechanics went outdoors. With the Loretta hidden in a jute sack, they traveled a good distance, to just outside the town of Nahariyya. They packed the shells with test weights rather than explosives, and tied red ribbons to them, to observe their trajectories.

"It was very nice," sighed Zvi. "One shell flew into a water tank and left a mark. I wonder if the mark is still there. Another got lost and we never found it, though we hunted all over the sands. Probably deep underneath someone's basement."

But the project that most excited Zvi had nothing to do with weapons. It was the equipment he and Itzik had built for diamond cutting, in which the kibbutz had briefly dabbled. Zvi loved pretty things. He had enjoyed the perfection required of the cut, and the exacting process of turning the dull, raw diamond into "a little glittering speck you can hardly see."

At two o'clock, Ziva and I went for a walk instead of napping. In the heat the birds had stopped singing, and we stepped out, with a pleasant sense of conspiracy, into a deep hush. We took a trail that led away from the residential hill, past storage sheds, into a grove of apricot trees. Under them, red sod had cracked apart in the shape of lightning bolts. Instead of rain, leaves' shadows slashed across it. Ziva looped a plastic bag over my arm, and we began plucking fruit. The crisp leaves thrashed against the sky's dazzle as I pulled the boughs, and everything smelled of dust and scorched sugar. The pink-freckled apricots dropped at the first tug. Ziva sprang up and grabbed a branch, one sneaker braced against the trunk, her wattled arms moving smoothly and quickly.

"I planted this grove," she said. "Plums too, real plums. Not what you buy in Jerusalem." Soon the bags were full and we were munching apricots, dashing stones from our lips.

"All the same," Ziva added, "I think the kibbutz failed Sharona. Somehow we failed to make her feel not alone, not hopeless, or how else can you explain that life meant too little? That she couldn't at least look around and say, 'What trees! What a sky!'"
I looked at the sky. We hung our bulging fruit bags from a limb, to wait till we returned from the graveyard.

A steep white track climbed a ridge with wild oats thick on its sides, their forked husks glinting and rustling in the wind, over the reverberation of crickets. The view gradually broadened to plowed fields in swells of brown, rose-brown, merging into a far lilac smoke. Each field's furrows were rounded in the corners; from the air, they would look like fingerprints. Ziva said the chief crop that year was safflower. The soil of the Negev was loess, a fine soil that had blown here over geological eons. Loess came from China. Our land is really Chinese, she laughed.

I caught sight of a tiny palm tree in the distance.

"Does that palm belong to the kibbutz?"

"It was planted for beauty, for the view," Ziva said. "Maybe it's not polite to say to a poet, but I believe that a field cultivated like these is equal to the highest of what is called art."

"You bet. I live in farm country."

"You must bring me up to date," she said, taking my arm. "I don't know where your life has passed. When Itzik died, I asked Sonia for your address and she didn't—"

"She had my address. Her lawyers had my address. The scientists also asked her, they didn't know where to find me. In the end, I found them." I looked back the way we'd come, down the white spine of the ridge. "It doesn't matter."

Ziva's sunglasses managed to look shocked. On her black lenses, my reflections shrugged.

"How can you say that?" she asked.

"I'm too happy now to care." It was true; I felt like a winner in the game of chance, the game of destiny, that had brought me to Israel in search of my father's past. We linked arms again, and I told Ziva about my father's second—no, third!—wife, and our peculiar relationship. Ziva took a provincial view: no Jewish woman, she declared, with a Jewish mother's heart, could have mistreated a child so. Like a witch, a Russian Baba Yaga, she said scornfully. I did not think Sonia's problem was her lack of Jewishness. I changed the subject, and we went on talking until Ziva asked where I'd gotten my fluent Hebrew, and I told her that between the ages of eleven and thirteen I had lived, with my mother, in Jerusalem.

"No," she said, halting. "I cannot believe it. I cannot—can you think of a reason, any reason, why Itzik would not have told me you were in Israel?"

"Probably didn't occur to him."

She was really upset. "You could have visited, with your mother too. You could have used the swimming pool."

I remember the beautiful, mournful way she said "swimming pool," as if it were the best of everything.

We reached an iron gate. A row of cypresses, pitch-black, were flailing and groaning in the forceful wind at the top of the ridge, where the sky felt closer than the earth. Ziva unlatched the gate and we entered a dimness. "Forgive me, Ziva. . . . Why did you and my father split?"

Still holding my arm, Ziva guided me down the double line of graves, which looked like stone beds with rosemary blankets. The pillows were white slabs, with carved names. Matted pine needles crunched under our feet; overhead, a roof of cone-studded, tangled limbs. Spots of rainbowy light quivered on the pale graves; and the clusters of pine needles cast furry shadows. The wind's noise was muffled.

"There's my best friend," Ziva said tartly.

Off to the side was a wooden bench. We sat, stretching our tired feet. Ziva said quickly, "To answer you: I discovered he had relations elsewhere." She sighed, and went on with practiced lightness. "I was too young, and too stupid, to comprehend his nature, and that it would not change. Itzik had a large appetite. He required plenty of room, lots of women, he had no use for the bonds of marriage. . . . It was his nature. I don't hold it against his memory."

"So . . . that was that?" Toying with the sunglasses in her lap, Ziva smiled her unequal smile.

"I told him," she said, "one bright day—'Itzik,' I said, 'let's throw away the whole loaf instead of eating the shit slice by slice.'"

"Oowah!"

"Now, of course," she ended with her radio crackle, "I am

experienced enough to know that in life, one throws away the loaf and eats it too!"

Sharona's headstone was the last in the row, which made it the first in the short history of the kibbutz. I bent down and put the customary pebble on the grave. It was the only pebble. I felt sorry for the girl whose rosemary blanket had withered, leaving bare, dirt-clotted roots. It was queer to see my own name staring up from a headstone. Was Ziva right, had my father named me out of an exile's wishfulness for this spot of earth, where his comrades were digging their first, lonely grave? I wondered. The wind was surging in the pines. I felt aloof from my namesake, unable to sorrow for her. Though this was itself akin to sorrow; it was what she had failed to learn, the toughness that diminishes feeling, the wryness that replaces it. I knew she had been in agony.

"Like the world isn't full of boys," Ziva scolded, more cross than one usually is with the dead.

"All the same, Ziva . . ."

"Mm."

"All the same," I blurted, "I don't think my father loved me. Or why would he have named me for a suicide?"

"But I think Itzik didn't love me so much." Her voice was altered completely by a slight shake. "Or he might have named you after me?" We hugged, though neither of us shed tears.

That night, the founders of Shoval gathered on Ziva's porch to crack sunflower seeds and reminisce.

They said:

Itzik astonished us by saying that someday tin buckets would be made of plastic. Who dreamed that he'd go to America and make a million? He was a good worker—beat the other men at hauling cement hods. True, but Itzik was *yefei nefesh* ("beautiful-souled," i.e., squeamish, a Marxist insult). Remember when we could only afford milk for pregnant women? He craved milk, spoke of it constantly. We went to a comrade who was breast-feeding. A bit of a joke, yes, but milk is milk. We brought the cup to Itzik, who drank, smacked his lips, thanked us gratefully, and when told whom it came from, turned green and spewed his guts out. He wasn't a man for daily wear. Itzik? Bentov? The boy was an *astronaut totali!* In the middle of the night on my way to the cowshed I'd see the lamp burning in his tent, where he was read-ing his engineering books. But he was so comical, *wal'la,* he could make a stone laugh—at classical music concerts, it was risky to sit next to him. In '78, he came here and told everyone that Mahananda the Indian yogi was teaching him to fly.

"He talked about you on that visit."

"About me?"

"Yes, he said he paid too much in child support. Oh!—but right after—I remember, he said it was worth it."

Finally a comment was received, by all, in the spirit of the last and best word: "One of us remembers this, another that, but what we all remember is his laughter."

Here are two letters, one from Moshe (in his English), and one from Ziva (in my translation). I received these in response to some questions I had sent them.

I asked Moshe about history; Kibbutz Shoval was one of the original Eleven Points, eleven wall-and-watchtower kibbutzim created to prevent the British from claiming the Negev. Moshe answered my questions by number.

1. *Daily Life on the Kibbutz in the Beginning:* Well, I was most of the time
until the end of May 1948 in charge of our carpentry shop. In Shoval we had
only a small core of 30–40 members. Part were engaged in laying the water
pipe lines or worked on constructing the Water tower and Security House.
Besides this there was of course a lot of night watch, of cooking, washing, etc.
But there was also the first optimistic beginnings of agriculture. In November 1946 we hired a tractor and drill to saw our fields with wheat and barley.
When the first rain came down our fields changed color from brown to
green. The young grain attracted thousands of astonished birds. You asked
what did the Chawerim [comrades] do in Shoval in those first days? Well,
they were recruited to defend the young and helpless plants from their winged
enemy—the birds. . . . After the early November rains there was no more
rain. What had been left by the birds in our fields was killed by the Draught.

Here in Shoval we were lucky to have good healthy loess soil. So we
needed not to wash salt out of our soil. But our neighbor, Kibbutz Chatzerim,
west of Be'er Sheva, was settled on salty soil. They struggled hard to cure the
soil and to wash the salt out of it.

2. *The U.N. Partition Decision:* At that night the whole kibbutz was gathered around our radio in the Dining Room. We counted the various countries
who voted pro or contra the partition of Palestine. When the final decision
was announced, all political objections were wholly and totally forgotten. A
spontaneous enthusiastic Hora swirled wildly in the middle of our tiny kibbutz. Three circles of excited young optimistic boys and girls. As far as I can
remember Itzik Bentov was participating in this unforgettable Hora. . . .
This was the last evening of peace, the next day the Arab riots started.

3. *Life on the Kibbutz During the War:* To prepare ourselves towards attacks
by Arab marauders, against the Egyptian and Trans-Jordanian armies, their
planes, tanks and artillery, we started to "fortify" the kibbutz. We build 4
"Pillboxes" in the 4 corners of the Kibbutz, digged trenches and air raid shel-

*ters and build a dam of earth all around the Kibbutz. We placed a few pipes
into this earthen wall that from afar looked like camoflaged artillery. . . .*

*To get building material for the air raid shelters we dismantled aban-
doned earthen baikot (houses built by beduins). Out of the roofs we took
rails (Krupp 1897) the beduins had taken from the dismantled Turkish
railway from Lydda to Be'er Sheva. We also took from there much Turkish
pine wood that had served under those rails. . . .*

*During late September and October 1948 Ziva and I were recruited to
a stronghold 10 km south of Faluja. Now the Egyptian Army was already
encircled in Faluja. Our task, together with the other recruited kibbutzni-
kim on other hills was: to avoid any breakthrough of the Egyptian Army
out of the trap that had been closed on it. . . .*

*But we did not only engage in fighting and military actions. Whenever
possible we used our agricultural equipment (tractor and tractor driven
Allis-Chalmers combine harvester) to cut and thresh Wheat and barley
all over the Negev. . . .*

Ziva wrote:

*. . . I bear no resentment for what was; moreover, my life has developed to
my taste. I would never have been able to endure a faithless spouse (according to the ideas of those times, and my ideas today, though they be old-
fashioned). Itzik had a Greek perspective: he valued aesthetic beauty. I
had and still have a Jewish ethos: I value beauty of the body-soul-spirit.
(Meanwhile, I have met on several occasions women to whom it was
granted to love Itzik.) I desired very much to meet you, the daughter of
Itzik. And your mother. That was after he no longer lived. Friends took me
to the address of his house near Boston. No one was home. His last wife
sent me his book, and a book of her sculptures. . . . When you came to
Shoval, Sharona, I was very very happy that Itzik left us a thing as fair and
rich and singular as yourself, with affection for this land which was and
remains my greatest love.*

Six

FIRST INSTRUMENT THAT'S FULLY CONTROLLABLE
BY DOCTOR IN PROBING THE CIRCULATORY
SYSTEM IS NOW BEING DELIVERED
IN PRODUCTION QUANTITIES. . . .

A new steerable angiography catheter, guided by a doctor's thumb on a miniature joystick, will soon take much of the tedium and hazard out of hospital procedure for probing the human circulatory system.

The new instrument . . . is expected to expand the physician's ability to examine, diagnose, and treat areas of the body cavity that are now not accessible to him except through surgery. . . .

Rapid development. Itzhak Bentov . . . designed the catheter in his crowded home workshop—possibly the only such shop to include a working plastics extruder along with the usual assortment of machine tools. . . . He started work in 1967, when two radiologists from a Boston hospital asked him if he could design a catheter with a tip that was steerable from outside the patient's body . . . he returned three days later with a working prototype. . . .

—*John Kolb, in* Product Engineering, *1969*

L ET THIS BE the invention that stands for my father's achievement. Not the rocket or the Loretta, built to kill, and for God's sake, not the Transcendometer, which was worse.

At the time this article was written, my mother and I were still living in Jerusalem. When we returned, my father, newly prosperous thanks to his catheter, had moved out of his first house with its basement laboratory, the "crowded home workshop" that impressed the article's writer, into an affluent suburban home. Why my mother and I went to Israel for my eleventh through thirteenth years is a mite puzzling. When I ask my mother, she says it was my father's idea. He'd convinced her that I would never speak Hebrew fluently unless we moved to Israel right away. I did attend Hebrew school, but it was felt that a native fluency would—what? What vital advantage was a command of vernacular Hebrew supposed to bestow?

"Your father said, 'Do it for Sharona,'" my mother explains. "My friends thought I was crazy. 'What?!'" she mimics. "'Interrupt your dissertation, and go alone with a child to a foreign country?!' But you remember your father," she appeals. "He could be very persuasive." Perhaps like me—like the ruby laser ray that was split in two, to make a hologram—my mother was split at some point into separate selves, one bouncing out into the world, the other unable to account for herself.

As for my father, I think he wanted us out of (what was left of) his hair.

In our absence, our apartment was sublet to three college boys, who stole, smashed things, and attracted roaches, ants, and fleas. Our landlord would not renew the lease, and this was a loss. Because finding an apartment during my childhood, I can tell you, was no easy business. Elderly landladies with powdered dewlaps shook their heads and said they wanted a family, dear, not a single girl; while male landlords said "divorcee" with a smile full of repellent considerations for my mother, and a suspicious glance at me. Returning from Israel, then, to find our belongings filthy and broken, and our home gone, we moved in with my grandparents, who always housed us temporarily when rent rose or a job ran out. Grandpa, a tempestuous baker, and my volatile mummy, who shared an impassioned rage at each other, fought tooth and nail through every one of Grandma's savory dinners, their underlips baring an identical line of teeth, like President Nixon's. I would eat with one hand, leafing through a paperback; the quarrel would stop long enough for Grandpa to growl, correcting my manners, "Sharona, put down that book."

Admittedly, when I was reunited with my father, he said, "Your Hebrew has improved one hundred percent."

Our Saturdays resumed. I was thirteen. My father drove me to his new house in Wayland, an upscale suburb. The road wound deeper and deeper into a realm of leaves, interrupted only by white mailboxes. His house was in a wooded gorge, and he cursed in Russian as he tapped the brakes, going down the steep driveway. I had never seen an automatic garage door before; I sank further out of my depth as we strolled around the house, my father pointing out the deck that hung within a stone's throw of the woods. On the front door, a brass dolphin knocker, pure Yankee, made me deathly unsure of my manners.

"Ben?" Sonia called as we entered. "Oh," she added. "Hi." She looked down from the top of the stairs, which split at the entrance into short flights up and down, carpeted in gray. The top floor was suspended above the stairs. As we climbed, a kitchen opened to the left; Sonia stood at the sink with her back turned.

"Lunch is ready," she said. It was a trap. The other time I had dined with Sonia lurked in the carefully forgotten past; all that remained was a foggy glimpse of my chair having stood, somehow, not at their table but in a doorway behind it. I sat down sick with tension at the butcher-block table, across from my father, who was nervous too, because he began selling me on the lunch to come.

"Yah, Sonia's special borscht! The best stuff in the world!" He smacked his thin lips as Sonia wheeled in a cart. My gaze dropped. I stared at the white plate wailing in my mother's voice: . . . *and the Royal Copenhagen wedding china he took when I was in so much pain and so much shock I just told him to take what he wanted and he exploited me, he grabbed that china my mother bought for us, he should never have taken it, it was indecent.* My father supported a Swedish rye cracker on his fingertips, gently daubing it with anchovy paste. As he nibbled, he motioned with his eyes toward the platter of cheese and spreads, appetizing, impossible. The crackers would crack. I would clutch, while they both looked on, at oozing, crumbling, mortifying clots. Safer to go hungry.

Sonia sat at the head of the table, her shoulders swiveled toward my father. Her waist had thickened, which was dimly endearing; it meant that she was real beyond my fear. She chattered and tossed her short, graceful curls like a round-faced schoolgirl; one incisor pressed into her pink-frosted underlip. She was feminine success incarnate. I felt uglier than usual, especially when her playful rolling eyes jerked, or flicked, away from my corner as though magnetically repelled.

The conversation offered no openings. Sonia told a story involving many names, Russian phrases, and an art gallery, while her husband kept up a sage accompaniment of "*Znayu,*" "Yah," and "*Nu*, obviously." Once he caught my eye. He crinkled his nose and called, "Hey Pookli!" in a cajoling tone. Sonia whirled in her seat; I ducked my head. Then she scraped her chair backwards. She stood up, bent over, and began to ladle out the borscht. A bowl hovered into view, settling under my nose. The soup it held was cold, the consistency of water, and the color of blood. It scared me so badly that my spoon rattled against the rim. One splash. One crimson-spreading splash would be all it took, just one, for me to be eternally disgraced.

"Oh, this is delicious," I quavered.

When, at last, I risked shifting concentration from my spoon, I heard Sonia saying, almost devoutly, the word "humor."

She and my father were praising a new Hollywood movie about karma. I was puzzled. My mother had raised a cinema snob; my favorite film was *The Bicycle Thief.* But my father and his wife were sold on their romantic comedy—they would go again. They advised me to go. Sonia looked me straight in the face and said, "It is very light, of course, but you'll see, they are telling the truth about spiritual evolution in a way that the common people can understand." Her undertone hinted at something distasteful.

"Great!" My father flourished his rye toast. "They are 'telling it like it is'!"

Gosh, I wished he wouldn't try to sound hip.

Lunch finished, he and I ran down the carpeted stairs into his new laboratory, which occupied the ground level. Gone was the cluttered basement of my childhood. He had so much space! A window spanned the area, a ribbon of sunlight and birch trees. With a springing tread, my father led me to a cleared worktable.

"Here she is, hey Pookli? Pretty damn good, in my 'umble hopinion," he mugged. He held a brass cone in one hand, showing me a brass-knobbed joystick set in the cone's base. Smooth in his movements, he was perfectly focused, purely himself, so handsome it hurt to look at him. "Now you try it." I was afraid to, I'd barely managed the soup spoon. But he insisted.

"You cannot break it. I made it very sturdy, to withstand the stupid doctors." From the cone's nose, a thin plastic-coated cable sprawled in loops over the table. Thumbing the joystick, he made the last four inches of it curl, straighten, then whirl in a circle.

"Three hundred and sixty degrees," he pronounced, not quite crowing. He explained that there were four wires in the cable; they relieved one another's stresses and supported one another's reach. "Now you try."

I shook my head. Patiently, he laid down the catheter and opened a foam box. Inside, under a protective dome, lay a human heart rendered transparent, down to the thinnest of its angel-hair capillaries. I cried aloud at the sight.

"I had to wait for someone to die." He rolled his r's gleefully. "The rright guy to make it legal. Is not permitted to take casts of the human heart without consent." I musn't touch, he warned, because finger oils stained the latex, but he urged again that I try the catheter. He had tested it on this very model.

Gingerly I took the brass cone with both hands, while my father inserted the tip of the cable into the aorta, its broad arch topped with glassy chopped-off arteries. I thumbed the joystick knob. The cable tip struck spiritedly at the vessel wall, and I sucked air through my teeth.

"What if it pokes a hole?"

"It cannot. Try to exert control. Better, better . . . now direct it into the ventricle. Try to avoid contact with the valve."

Concentrating, I nudged the brass knob with the fine incremental touch I'd learned on my microscope. The probe, blurred under the clear tangle of coronary vessels, lengthened bit by bit, creeping into the tapered chamber. There inside the valentine's point it curled up, like a curious worm, bowing this way and that.

"Good," said my father in profound contentment, over my shoulder. "Even a little girl can operate this device." Seeing my face, he added quickly, "A big little girl. We can use this to inject photo-optic fluid too."

"What's that?"

"It is a fluid injected into the bloodstream." He said that it caused the circulatory system to light up, like a glowing diagram right through the skin, when viewed with the proper instruments. A dangerous blood clot showed as a dark blot, and you could send the catheter inside to pierce it, letting the blood flow freely again.

I laid the brass cone, full of delicate imponderables, back in its foam box while my father talked on, happy and proud. In a stray sunbeam, perhaps the ghost of Dr. Weizmann paused beside us, smiling gravely as my father boasted, "People will be saved."

Seven

HAMLET WAS RIGHT: I can't tell if he's offering me an illusion, or a true vision. Maybe that's in the nature of fathers. Sooner or later, I must decide. I'm sitting in the glass-walled sun room (on this rainy day, the cloud room) watching my father on a videotaped TV show from 1977. He was losing his mind then, at the height of his public success.

The show host is a Unitarian minister; young, pink, with a bucktoothed smile, silky hair fringing over the notched collar of his white leisure suit. His name is Stan. He introduces Ben Bentov as the author of a book that's a best seller among people interested in the relationship between consciousness and the cosmos.

"Ben is also a loving, wonderful man—"

Interrupted by my father's off-camera giggle. My father is himself enough to walk out of the TV. I'm impressed, at my age, at how his shirt barely pouches between his shoulders and belted waistline. He's dressed Israeli style, short sleeves, open collar, chest hair in the V. But he's not well, clearing his throat and wetting his lips. The studio basks in the buttery light of a morning show, and the men sit in plastic chairs, one rubber plant beside each.

"Can ordinary people like me," asks Stan, making dancy host motions with his fingers, "improve their minds through meditation?"

As the camera circles back to my father, it catches him in the act of leaning, back to the audience, over the side of his chair. He straightens up. Twice, he repeats the maneuver between question and answer. Twice he shields some furtive mission from the audience and sits back up, foiled.

"Yes, of course," he answers. Now the camera pans back to Stan; but as my father dives for whatever lies beside the steel pedestal of his chair, the camera nabs him again. He sits back up. "Improving the mind is only the beginning toward an overall refinement of the nervous system." He is plainly miffed.

Long ago, he told me it was a bottle in a paper bag.

"Cough syrup, Pookli, and the moron cameraman keeps following me! So the audience can think I am drinking viskey like a bum. *Tchort!*"

My father rises and walks over to a flip chart. He looks wearily around for the marker, finds it. He draws a bell curve, reassures us—"Don't vorry, is not very scientific"—and explains how most of the population fits under the bell with a few on its rim. To the far left, stupid people, "A gorrrilla type who beats his chest when he sees his neighbor." Under the bell, "The bulk of the population today, an intelligent—more or less intelligent [laugh]—biped, who pushes the buttons on the TV."

I'm amazed at how these condescending quips are changed by his delivery. Stan and I are both hunched forward, happy as clams, feeling loved. Loved! It's his charm. He seldom looks straight at you; when he does it's not with a direct gaze but shyly peering. His masculine, almost gothic, face has the wiles of a sweet little boy's.

. . .

By now, he had solved the problem he used to talk about on our canoe trips: how the human mind connects with the giant hologram composed of all the wave energies in the universe. He had built a complete cosmology, drawn from his researches, his general scientific background, and the popularized Eastern religious ideas of the time. It started with the ballistics of the heart: how blood, shooting from the left ventricle, makes the human body vibrate. (He knew the heart from his life-saving cardiac catheter. He knew ballistics from Israel's first rocket. Thus his main idea sprang from life and death.) His book was called *Stalking the Wild Pendulum: On the Mechanics of Consciousness*. And if it led him unwittingly to do the devil's work, that is still no reason why I shouldn't explain his two-part system. Your pendulum, then your consciousness.

Don't worry, it's not very scientific.

A pendulum. Say, the round brass bob inside my mantel clock. It swings up, rests for a split second, swings back. To get to the rest point, it crosses shorter and shorter distances—but below Planck's distance, so short only physicists bother with it, something bizarre happens. Time and space separate like curdled milk, and become chunky. A particle might cross a chunk of space without a matching chunk of time. The bob (as the sum of its particles) actually escapes time, crossing space chunks, traveling those eensy distances in, literally, no time at all. This must happen at infinite velocity—i.e., faster than light. A pendulum bob traveling faster than light. Inside my clock. On my mantel. Can this be?

My father cited Heisenberg's uncertainty principle, which says that you cannot know both the position and the momentum of a particle. If you know one, the other turns indefinite. At the rest point, we know the pendulum bob's momentum is zero. So its position turns indefinite. You can't tell where it is! It's in all possible positions, anywhere, everywhere—"even at the end of the universe." (My daddy! If he needed a special tool, he built it; if he needed the end of the universe, he dragged it in.) At the rest point, the bob expands—at infinite velocity—to all possible points, like a sudden blur over the universe. Then it starts swinging back, calm and small and round again, and my clock serenely chimes the hour.

Atoms, which vibrate periodically, are tiny pendulums. So atoms do this too. He wrote, "[I]t is very likely that our matter is blinking on and off," each atom blurring over the universe, then shrinking into itself, on, off, on, off, like fireflies in a dark wood. Our very bodies, vibrating up and down to the beat of our hearts, are pendulums. They must expand to the ends of the universe and contract again, roughly fourteen times a second.

Nothing my father ever imagined was restful.

That is the wild pendulum. (If you think he extrapolated incorrectly from quantum to large-scale phenomena, send me a letter—a very very short one, really really fast.)

Now for consciousness and its mechanics.

Remember, he wanted to know how his mind could make contact with the giant hologram composed of the universe's vibrations. As the body blurs to all points of the universe, it's in contact with the vibrations of all nature, including those produced by thinking brains. But it can't register them. He needed an "observer," or a soul. The soul, he wrote, is a nonphysical "information-processing unit" that stays aware, while the body blurs across space. Normally, we can't remember the event; our nervous systems don't work that fast. But in altered states of consciousness, like deep meditation, our sense of time slows; and the "observer" has enough slowed-down, subjective time to register, and recall, some of the amazing information in the universal hologram. "All creation is in constant and instantaneous contact," he wrote, "with some creatures being more aware of this than others." Thus some people could receive information—sudden illuminating ideas, flashes of inspiration—which came from their contact with the giant hologram containing all the data in the universe.

To answer Stan's question, meditation didn't just improve your mind, it helped you evolve into a being with a more skillful "observer," a *higher consciousness.*

Those two words? They still make me sick.

Okay. Consciousness—as he defined it—is the capacity of a system to respond to its environment. Beam ultraviolet light at an atom, for instance, and it will respond with excited electrons. That is its way of being conscious. By this definition, all matter is conscious.

Now, my father never read any philosophy; as far as I can tell, his notion of what went on in that discipline amounted to far less than his information about, say, double-threaded screws. So there were questions he never asked. He never asked if consciousness entailed some form of self-awareness, and what that might mean. He never questioned, in his definition, the role of the experimenter who beams light at atoms—he didn't think twice about using the context of empirical science to frame something as philosophically elusive as consciousness. He drew no line between living and nonliving matter. According to my favorite biology book, *What Is Life?* by Lynn Margulis and Dorion Sagan, you can't just go around imputing consciousness to everything plus the kitchen sink; that's animism—the belief that everything has a soul, which I happen, personally, to be very fond of, but it's unscientific. There are wondrous, seductive paths of scientific reasoning that converge on a definition of consciousness in living beings; as I understand, there are no such paths for the dead dust we're made of.

Yet, credit where it's due: my father's animism was pleasingly up-to-date. He thought in terms of multilevel processes, as systems theory had begun to teach people to do; he defined consciousness as a process of response to the environment. The nutshell version of his ideas was that *matter is consciousness.* Where there is being—be it the least mote of dust—there is the consciousness of being. This would mean that the received modern Western distinction between conscious mind and its opposite—unconscious, mindless matter—is an error. Matter is mindful. Matter minds. In Act 4, scene 1 of *The Tempest,* the wizard Prospero admonishes us that "We are such stuff / as dreams are made on." We are accustomed to distinguishing objective reality from subjective reality—that is, objective facts from subjective feelings. But if "matter" is "consciousness"—if these are two names for the same idea, only with different inflections—then it follows that "objective reality" and "subjective reality" are also names for the same idea. There is no strict division between a fact and a dream. Facts may be dreams, and dreams, facts; imagination, from which dreams arise and by which they are shaped, may be the foundation of what we call the real world.

Such notions are far from new—you can find them around; but to me, they are as comforting as hot cocoa on a snowy night, or the first crocus spearing through a crust of melting snow. They are my father's teachings, the faith I was raised in. Yet I have reason to suspect that it is a mad faith. If I don't get to the bottom of what makes the difference between a true vision and an illusion, I will have no peace.

My father began to go wrong the way engineers who want to improve humanity usually do. He started measuring. He decided to measure consciousness. How do you measure consciousness? By quantity, the number of possible responses. And quality, the complexity of responses. He ranked beings on a curve from lower to higher, from the less evolved consciousness to the more evolved. Rocks were lower than plants. Plants were lower than animals. Animals were lower than guess who. The scale was uninformed and kind of silly. Find me a botanist who'll agree that plants respond less complexly than animals! As for human beings ranking higher than other species, that conceit is finally taking the flack it deserves; for all my fancy brains, ask me to respond with my naked nose to the environment, ask me to conduct a urinalysis with this organ, and I lag hopelessly behind my dog. The evolution of consciousness, said my father, required the nervous system to be refined. Well, this was a man who had left his kibbutz. He thought of beings as lone individuals, preferably with brains. Bacteria have no brains at all, no nervous systems to refine. Yet it is hard to imagine more complex, and immense, responses to the environment than aerobic respiration and photosynthesis, which regulate the oxygen in our atmosphere, and were originally created by huge communities of bacteria, life's pioneers, working in concert the way they do. Damned if I'm going to measure my speck of consciousness against the Himalayan collective soul of the earth's bacteria! Charles Darwin once noted that one should never employ the terms "higher" and "lower." Hierarchies of values exist only in our minds, not in the biosphere, and evolution is not the factory conveyer belt my father imagined it to be, making progress toward a predestined end. Evolution is a process governed in the main by random chance, that spark at the root of creation.

None of this is what sickens me. It's fun to find your dad's mistakes, harmless errors of theory. The evil, the really bad medicine . . . is another story, for later.

For now, planted in front of the TV, I watch, with a surge of bone-deep yearning, my father draw his bell curve of lower and higher consciousness. Because now he is going to tell the Sunday viewing audience about the vision I so long to keep. Once upon a time, I was a wizard's daughter; I want it, him, back.

His dark-haired, muscular arm reaches across the flip chart; he marks the bell curve, here and here. What we perceive, with our senses, is our reality. But highly evolved human beings, whose consciousness is higher than normal, perceive realities that are, to other people, invisible. You have heard of people who see angels or spirits? These are the nonphysical realities. We live in more realities than we think. We live in many realities, in multiple dimensions, but they are invisible unless we learn to perceive them.

Stan and I are transfixed. My father is revealing invisible wings attached to our shoulders that can lift us into worlds beyond, where the dreams that you dare to dream really do come true.

Of course, these unseen realities are also ranked, lower to higher. The lowest is astral reality. It is the level of ghosts.

"Are ghosts real?" Stan worries off-camera.

My father laughs. There are no dead people, only people

without physical bodies. Ghosts who slam doors or break crockery are slumming; they can affect physical reality because they've stooped to our level. People who see ghosts, on the other hand, are boosting their consciousness to a higher level. There are techniques, yogic techniques, that can push your nervous system to function in the higher realities. . . .

He begins pacing with genuine, if hoarse, enthusiasm before the drawing on the flip chart; he smiles at it, taps its highlights. The ultimate goal of our evolution is for all human souls to merge in the highest reality of all, the mind of God, or Cosmic Consciousness.

"Pyoorr consciousness," croons my father, in love with God.

"Shit," I mutter. I hit pause on the video remote, and sicken at the sight of my father's image—jarred, pixilated, stuck.

The hardest task is separating the Promethean flame of creativity from the fires of madness. The trouble is, people like Uri Gil, respected scientists, listened to him—smilingly, but they did, if only for the original bent of his mind. For me, growing up, there was also the problem that success is hard to question, and he was idolized by thousands of New Age followers. They were not hippies, not edge-walkers of any sort, but upper-middle-class professionals, who packed his lectures at luxurious retreats and international institutes; they were so much more respectable than my mother and me. Who was I to doubt? I envied them. I had fewer Saturdays with my father because of his packed schedule.

I lost the faith I'd been raised in at his memorial service, held in a modern, well-lit civic auditorium with subdued carpeting and good acoustics. Hundreds of people attended, it seemed, as I looked around, folding my invitation into origami. In front of the stage was a bank of yellow chrysanthemums from which rose a pole, on which, impaled, staring out at the assembly, was a life-size bronze of my father's head. I kept waiting for his expression to change. The blind bronze eyes mirrored my own appallment. People trotted onstage and spoke. Stan was there, to remind us that there were no dead people, only people without bodies, and to suggest that Ben was with us now, "laughing." A pleased murmur ran through my father's friends; they could see him whooshing around the auditorium like a Disney ghost, pulling funny elastic faces. Another disciple trudged up and said with gloomy optimism, "Ben has skipped the body." Meaning that my father had voluntarily got himself exploded, or, in lay terms, committed suicide, because physical reality, where I was, wasn't good enough. One blithe spirit in crocheted flax hinted that the cosmos itself had promoted Ben a step closer to godhood.

As I listened to the ruthlessly upbeat eulogies, I began to rage. In loud unspoken terms I told those people that they were gutless fools, that pain and death were not explainable, not reasonable, not in their control. If my father's ideas were responsible for this dross dished up in place of honest grief, then I hated his ideas, I despised his teachings, once and for all I scorned, as worthless illusion, the faith I had been raised in. But the rage was superficial, and its venting, in scowls and mental oaths, brought me no ease. It was—the rage was—no more than the heat from a ball of fire that grumbled on the floor of my mind. Inside that fireball, my handsome father burned to death. He was alone. I was not near him. This would not stop while I lived. My one solace, which I clung to, was the pain of knowing that this had been real, and nothing could lessen it. I grasped hard at that branding truth, which alone was worthy of him. My father had died beyond the scope of human reason, beyond the fancies of cowards and fools, and the best guesses of the wise. He had died by chance. By stinking, lousy luck.

Am I so different from my father's disciples? Haven't I never quite relinquished the hope, snugged in deep, that my wizard father can do anything? I don't really believe he's gone. Physically dead, all right. Gone? From all the realities? The story's main character?

I restart the videotape. The camera pans back to the rubber plants and the show's host.

"Ben, can you tell me, tell us, are there actually people whose nervous systems have evolved to the point you're talking about, where they're actually superhuman beings? And if so, what would they look like?"

My father drops in slow motion back into his seat, looking haggard.

"Vell," he says, and coughs. "Suppose you tell me. Where do you think we are finding the most spiritually evolved people?"

"Ah, in universities—they would be intellectuals, our great artists, uh, Nobel Prize–winning scientists . . ."

My father's lids have sunk over a spark of glee. His broad chest is caved in the contoured chair as though he'll never rise from it.

"No, no," he murmurs. "They are sitting in mental institutions."

Eight

ROCKET MAN! The man who assisted my late father in building Israel's first rocket. In my little photograph of Hemmed, I recognize the shape of his face, a white grape with ears; a slim, smooth, innocent face, lifted on a tender neck. He's twentyish but looks fresh from his bar mitzvah, kneeling on one knee, hands folded, blank with solemnity, beside preoccupied Shai.

Such a cute boy, I want to give his cheek a motherly pinch, there, on that fleck of white. He's nothing like his older self. Here is a description of Misha in my Israeli notepad, written the night I met him. Uri and Sara Gil had invited us both to dinner so I could hear the story that my father never told me.

My father's assistant! Short, slender, seventy, completely mercurial, silver crew cut seems to effervesce from his scalp, eyes are small, black, and brilliant. He and I hit it off instantly, he's voluble and funny, shy of the Gils—he keeps bubbling up in his chair, then looking abashed while they exchange their Long Married Look. Like he's the runt of the Hemmed litter. A retired nuclear engineer. M. has his own private story about Hemmed; he knows what he wants to remember, and how to remember it.

Yes, that was, that still is my problem. I wanted Misha to tell me stories that I could imagine, bursting with picturesque details, to bring my father and his world to life. Well, Misha was an engineer. He told me numbers, percentages, diameters, rates of fire. He told me his political opinions, and a few bald anecdotes, not in any particular order. We kept getting sidetracked. Once I asked (subtly, I thought), "When you first saw my father, in the Gray House, what impression did he make?" Misha paused. He echoed the word "impression" as if it were the name of a curious, unwholesome hors d'oeuvre. Then he said, "An inventor."

An inventor was an inventor. X = X. X did not make "impressions."

Now as I review my notes, long on data, skimpy on life, I wonder how to fulfill the ancient rules of the Telling: each generation must tell the story of Passover (and, by extension, all the ancestral stories that make us who we are) as if we, personally, had made the passage from slavery to freedom. How to relate Misha's story as if I, personally, had lived it? It's borne in upon me, once again, that the Telling must be invented anew in every generation, perhaps differently for each story, by some— lucky fool.

I turn the clock back fifty-two years. There are no high-rises on the coastline of Tel Aviv. The beach is bare except for a diamond haze standing on the transparent sea, between the haunted gloom of Europe and the new city on the level sands.

Walls dazzle. Even the sandy gravel lining the streets dazzles in contrast to the newly laid asphalt. Sometimes a eucalyptus casts its swimming shade over the brown arms of a bicyclist who halts, astraddle, to pluck a newspaper from its clothespin on the display board, and toss a few piasters into the seller's cup. Buildings are raised on concrete stilts, elevating the city into one common arcade through which fresh air and citizens freely circulate. Tel Aviv believes in air and light! It is May, fifty-two years ago, and black spores vanish from the grout between the wall tiles of Café Cassit. Dull spots disappear from the chrome shanks of the door handles. Everything is new; everything shines, like the folded cream on today's cake, on its silver pedestal. The café door, standing open, reflects a curb that pedestrians cross like stained-glass versions of themselves, with a grace seen only in flattened reflections.

Israel is born. Where is everybody?

David Ben-Gurion, with narrowed eyes, is scanning a group of reporters because he has heard one of them call, "Your Excellency!" And he wonders who this potentate, this excellency, is. Until informed that it is himself.

Hemmed, on the roof of an actor's apartment, occupies a shed called the Studio—to mislead British spies, hunting for Hemmed's technical director, Yevgeny "Genka" Ratner. A gifted inventor who spent World War II in the British navy and created the limpet mine, he was denied permission to return home to Palestine at the war's end, to prevent his working with Jewish nationalists. But Genka tricked his senior officers; he faked a serious illness and begged to be allowed to die in the Holy Land. Sentimentally, they agreed. Now under cover as an artist, in a beret that on him looks military, Genka struggles with his recruits. One of them is the teenage son of the architect who put Tel Aviv on stilts. An architecture student who yearns to put

Bauhaus windows in Tel Aviv stairwells, he is now the Studio's draftsman. He jumps as the director's fist slams down on a weapons plan he is drawing, taking pains with the crosshatched shading, rounded volumes, subtle transitional elements, so it looks, for a three-inch mortar, really nice? Genka's jaws clamp; his nostrils work as he absorbs the total lack of standard drafting format in his assistant's charming sketch, due at the machine shop in fifteen minutes. Black-faced, baring his molars, he roars, "AAAARR—CHITECT!"

Inside the café, there's a sparkling music of forks and spoons. At a crowded table, the poet Alterman is expounding to his friends, a fellow artist licking cream off a knuckle, and the great theatrical diva, Hanna Rovina, affecting a lazy slouch. (My friend Commander No One has snatched time from his duties to come here, for the city of air and light is also a bastion of secrets, and none of his bohemian friends know what he really does. He plays the role of a young poet with secrets important only to himself.)

Alterman is praising a new production at the Dais Theater. With wise shakes of his finger, he says that the characters are wonderfully simple: a soldier, a kibbutznik. Not the sad effetes of Chekhov, nor the labyrinthine psyches of Ibsen, nor the kings and princesses of the Dais's biblical plays. They are Israelis, simple, of this hour, this street.

"They are ourselves," he says.

Rovina strokes an earring, by that gesture drawing the others' attention before they know it.

"We don't want to be 'ourselves,'" she objects in a girlish treble. "You know why? Because . . . you're twelve meters from the enemy, you see him, he's your age, he's shaking, his legs shake, he's scared spitless. And this gives you a bad feeling. A very uneasy feeling. You don't want to shoot him, you want to—to grab his arms and shout, 'Why are we doing this?!' I don't know if it's compassion, if it's right to call it that, anyway, something of the sort . . . Or there's . . . the sky. Then the clouds. Then the bunker roof. Yes? Down in the bunker, in the corner, like a speck of dust—there's you. That's how it feels to be 'ourselves.'"

She can lift her oval eyelids with the effect of a thunderclap, or the charm of a bride. She lifts them. Every other face belongs to a faded figure, stubbing out a cigarette or resettling a cup in its saucer.

"Act one, scene one," she says, her voice dropping to alto. "There's your soldier, it's a dandy part. But my dear friend! Are you calling that character simple?"

Outside the café, on top of a white truck belonging to the Tnuva Dairy Co-op, a seagull thrashes to a landing, flicks his tail, then returns to scrape the sky, glinting against clouds like a barber's razor. The dairy truck trundles down the glaring street and into a nursery rhyme about itself, which ends with the word "milk." It is spring.

—We had nothing. What we had was a joke! We had one book, an American manual called "Principles of Rocketry," and on the front cover was a big blue stamp: DECLASSIFIED.
—Why, Misha?
—The British made it illegal for Jews to own weapons.
—Why just one book?
—When the Brits left, they imposed an arms embargo.
—I don't get it.
—The embargo included information like books and blueprints—look, we

didn't have weapons. We couldn't get the information to make them. Your generation forgets. Plus, we were mostly students. I was still learning abstract principles. You know, maybe, to fix a broken lamp for your desk, right? And you're told to build rockets. You're sent to a secret place in Tel Aviv, with no materials and no information, and no experience.

—Was that when you met my father?

—I met him in April, he showed up in his big coat to fool the police. He found us through some rumor. Uri and I both received army orders, but your father and many others just showed up in Tel Aviv, where they would be adopted, so to say. We were so naive. To give you an example, one of our geniuses wanted a code name for his project, which was the infrared lantern. He called it Ultraviolet. To lead the enemy astray! That we managed to survive that war at all—you know the history? The British incited it, the bastards, their plan was that the Arabs would wipe us out, right? Then they would control the Arabs and the oil.

—Well—

—I was in a patrol on the Jordan; I saw Arabs on camels crossing over from Transjordan, at night, bringing weapons into the country under the noses of the British police. They counted on us being massacred.

—Well, but Misha . . . the British didn't invent Arab nationalism, and anyway, if there was a pro-Arab policy it was set by their Foreign Office, which was very different from the Colonial Office and from Parliament, as I understand. Not to mention English public opinion.

—For us, it was all the same bastards.

—My father used to cuss the Brits. After he died, I found that he owned a collection of imported English woolen shirts.

—Here he did not.

—Okay. Where does Hemmed fit in?

—In May the Brits leave. We declare the state of Israel. Then Egypt invades from the south, and Syria, Jordan, and so on. These are not local Arab militias, *these are national armies with tanks and fighter planes. I am getting to Hemmed! A few kibbutzim in the desert stood between the Egyptian army marching north and the rest of Israel. These groups, of fifteen to forty people, were holding out against artillery and air bombardment and Sherman tank brigades with nothing but a few submachine guns, if that, some mines— homemade, of course—and Molotov cocktails.*

—Didn't Ben-Gurion instruct them to hold their ground? Until the foreign arms deals came through?

—What, "instruct"! Sharona, sometimes it seems to me that the Arabs are playing a game. When a war is over they say, "All right, we take this many pieces and you take thus many." But in '48 we lost one percent of our population. One percent, it's like you Americans losing two and a half million people, not in a Vietnam somewhere but in your own home. It has a permanent effect.

—You were saying about Hemmed?

—The Egyptians came north . . . in miles . . . thirty miles, that's how close they were to Tel Aviv. That's what Hemmed was for. To stop tanks. Not to invent new weapons, just to copy what the Allies had already tested, so to say, during the world war. We knew that rockets, bazookas, flame-throwers all worked; it was a question of reinventing them without materials or know-how.

Don't say the new government lacks pomp. The prime minister's state-owned motorcar is the height of luxury. It is (1) a car, and (2) a *Krise-lehr.*

Most people do not own refrigerators. But at an ice-cream kiosk soldiers are milling around, a happy crowd sticking out their tongues, licking ice cream that tastes of zinc. Their shorts are rolled up to their hips, exposing the pockets. It's the fashion; air, light. Misha, Itzik, Uri, Shai, and the rest of the gang are not here. They might be in a place, the sort of place, that is, known to those who are intended to know it.

. . .

Let me try to describe the Rise. You must imagine a long, low sand hill, in an atmosphere of lazy, humid sea air that undulates like a wave. You think you see a dark spot here, a moment later you see it there. Sketched around this place, unearthly in its incompleteness, is a barbed-wire fence, and yes, a gate. At least, a guard. No, some sort of gate. Squint hard, and your tears might lense into sharp focus the sight of three low buildings. Each has a sign posted in front of it. Each sign matches the color of its building, and there are, as in a fairy tale, three: the Brown House, the Yellow House, and the Gray House. Whew—wipe your eyes. What you've seen is classified. Look somewhere else—over there, a stick figure and another stick figure are wiggling across the sand on their way to the parade ground, or the clinic, or the mess. The Rise is a professional, Israel Defense Force, army base, did you know? No more smuggling bullets in brassieres or grenades in the egg crates of Tnuva. No more briefings in cafés. We have a regular—if unpaid—national army, with ranks and insignia and everything. Even if officers routinely ask the troops for their opinion, even if we respect daring more than obedience, even if we violate military doctrine by making frontal assaults uphill and so forth—we are the real thing. We are on the Rise.

So what will come of the new insignia? Hemmed's idealistic Major General Katzir fears that official ranks could lead to *regimented thinking*. Hemmed is using snap-on insignia, adopting the rank of whoever is scheduled to meet with them. Snap! You're a major. Go tell Major Anxiety that we do not yet have rockets.

Which is what the Gray House . . . might or might not be for, I wouldn't know.

Shai spills the beans in his booklet *Hemmed Days,* published privately in 1996 and distributed to those who knew.

> The main business of the group of physicists in the Gray House was to establish a Ballistics Lab and check various weapons. In the lab, we designed and began to build three systems: a system to check the velocity of rifle bullets and shells; a device to check the rate of fire of automatic weapons; and a gauge to measure internal pressures of hulls and combustion chambers. . . . We investigated rocket-type weaponry. We read and heard about German, Russian, British, and American rockets. The recoilless cannon found favor in our eyes. We were interested in a light, mobile cannon. . . . An ordinary cannon, or mortar, shoots a light shell forward at a high velocity and the heavy cannon recoils backward. The recoilless cannon is a combination of an ordinary cannon that shoots a shell forward, and a rocket that produces a gaseous stream backward to cancel the recoil of the cannon. . . . In April 1948 two Technion students came to the Studio. [Misha,] an electrical engineer, and [another recruited science student] . . . at the same time as Itzik Bentov, a young kibbutznik from the Negev.

> —*Your father first came to the Studio wearing a big winter coat to fool the British police, who luckily didn't ask why he wore a coat like that in April, and under it he carried a small-caliber recoilless rifle he'd made on his kibbutz.*
> —*Misha, when you first saw him, what impression did he make on you?*
> —*"Impression"? . . . An inventor. Itzik designed the rocket and I assisted him. He used a water pipe.*
> —*Why a water pipe?*
> —*Because nobody owned cars! We weren't a rich nation like the United*

States with mountains of good steel sitting in every junkyard, right? For steel we had to use pipes. For fuel we used cordite.

—What's cordite like?

—Our cordite back then was stolen from the British, who used it in a gun called the twenty-five pounder. It came in sacks. Funny—I'd forgotten about this—it was extruded, it looked like macaroni! We called it macaroni! The pipe was fifteen inches long. Itzik plugged one end with a metal stopper and drilled a hole in it, for the fuse. If this design had even occurred to me, I wouldn't have known how to make a stopper. Itzik was ahead of us in practical experience, though he had no education. He was a mechanic, he'd fixed tractors. I admired that!

The sea turns maroon with sunset, and waves wash in like foaming rosé. Lightbulbs are swinging to and fro, strung from one beachfront café to the next, and music garbles out of competing loudspeakers. Between the cafés and the murmuring water two girls stroll by, fingers linked, skirts belling with air. A party of young married couples laughingly stamps their blanket onto the sand, weighting it down with a portable gramophone. An artist folds up her easel. A fisherman hauls nets on his bent back. There are no tourists.

Foreign journalists sit around suffering the noisy music of the cafés because if they stay indoors, it has to be in a stifling room with shutters and curtains closed, due to the blackout. After midnight they can return to the Kaete Dan Hotel and swelter over their midget Hermes typewriters, amid squashed cigarette stubs, watery Scotch, and black-ringed Turkish coffee cups, banging out reports. Bets, anyone? Which of these three gentlemen—the intrepid war correspondent, the debonair columnist, or the AP guy with his baseball cap—which one is about to scoop the Manhattan Project of the Jews? Like this:

It is the cradle of Israel's atomic development: the Science Corps. Prime Minister David Ben-Gurion is the father of this top-secret unit within the Israel Defense Force. In the past month, he has endowed a corps of leading scientists with a research budget which some sources claim amounts to . . .

Three thousand dollars. 3K. And I'm the one to tell you. Because whether they are sympathizers who think the Jews are courageous and touching, or antagonists who think the Jews are pushy and vulgar, the foreign journalists have inaugurated their fifty-year tradition of missing the point. They grumble about Tel Aviv. This town is dead, the only nightclub worth a damn is that Club Atomic.

At night, Israelis bend the blackout regulations to skulk about on their rooftops, watching the Egyptian air force show. *Feeeee-boom.* An apartment building subsides. Half a bus is crushed. Inside the jiggling Gray House there are shouts. Misha blocks a lighted window with his face, a perfect target. Nothing happens to him. He climbs back onto a stool and tucks his sandal under a rung, hooking his toes around the warm spot on the iron. Overhead, lightbulbs are swinging shadows round and round; the little room, cramped with pacing science students, bundled extension cords, free-floating papers, cigarette smoke, feels like it's tilting. Some experimental shells have rolled off the pile in the corner and lie wobbling on the floor, what a mess. He hunts for a fresh starting point in his mind, while the web of voices reweaves itself around him.

"Who has the logarithm table? I need it."

"SHO-sha-na, Shosha-na, Shosha-na! Dee DEE da-da-dum . . ."

"Is this today's coffee boiling here, or yesterday's? Is it after midnight?"

"—when R is the resistance of the circuit and T is the time it takes for a shell to travel n meters between the wired plates giving us n/T meters per second—"

"Crap! Not again!"

On a path that is no path, in gusting yellow dust, Ziva, muffled in a head scarf, runs hunched over to the row of sheds, and as fast as she can chalks four letters on each door, signifying that if this drill were a real defense of the kibbutz the whole row of sheds would now be EXPLODED EXPLODED EXPLODED EXPLODED.

In a ruined kibbutz, where no cement chunk is left upon another, where scattered nails, cow skulls, charred boards, here and there a person's bone, are mourned over by a dark, irregular eye blasted in the wall of the water tower, and a torn tarpaulin skids across the ground in the wind, as if searching restlessly . . . on a ruined barracks wall, some chalked graffiti slowly fades:

It is not the tank that will win. It is the man who will win.

Today, Hemmed is invited to the launching of a rocket built by an anonymous inventor in Israel's navy. Our friends pile into a jeep and drive to the beach at Herzliya. There on the empty dunes they encounter a vision: a rocket indeed, of large dimensions, resembling a sort of streamlined aeroplane. Three meters in total length and built completely of cast iron.

"This rocket," observes Hemmed, "is heavy."

Their navy host enthusiastically points out the engine inside the tail, where burning fuel will create the gas column for liftoff.

"What fuel?" asks Hemmed, propellants running through their heads.

"Kerosene," says the navy. Meanwhile two large Primuses, hand-cranked paraffin stoves on which every housewife cooks, does laundry, or heats bathwater, are being used to fire up the rocket, their greenish-blue flames spouting around the tail and through the engine aperture. Ignition does not proceed, however, for a long while.

That big iron body and wings are conducting the heat away, diffusing it, explains Hemmed. The navy shrugs.

At this moment, dripping kerosene at the aperture ignites, and the rocket lets out a ground-shaking roar. Geysering greenish-blue fire towers. The big rocket gradually glows red. And while the rocket scientists watch, it stands as unmoved as the pyramids.

In the dead of night, Misha hears himself agreeing with the others that the internal pressure gauge does not work because the spring inside the hull is too stiff to register combustion because because because because because.

"Our design specs were correct," Misha croaks.

"Yah," says Itzik. "But the reality was wrong."

Then Misha forgets how his eyes are designed to open.

The army tries to find veterans of the world war who have had contact with rockets. Two are found, both Russians. One has painted a Katyusha. The other, once, fired a Katyusha.

The Gray House ballistics lab extracts what it can from these accounts, and keeps on working through the watches of the night, and through the hour when blue separates from gray, on through the hour when a breeze flows inside smelling as if thistles had been fermented, dried, then drenched with dew.

Misha yawns extravagantly, thinking it smells like dawn, it's dawn again.

A white-haired farmer in soiled overalls, arms held away from his sides, enters a bunker. He stops for a moment, his fists rise, then jerk down as if yanking invisible ropes. There is reddish light in the bunker, a typewriter and wireless set on a table, three rifles in the corner. A big field officer looks around and interrupts the farmer, who has begun to shout.

"Was there a vote or not?" snaps the officer. "Two weeks ago we voted to evacuate the children and to stand firm. I have a lot to do."

"Two weeks ago! Before the Flood!" roars the old man, yanking his ropes to his knees. "The army experts come, they talk, the stupid world salutes! I want to know why eighty good cows, cows that cost me fifteen years of hard work! Why weren't *they* evacuated? Why did those planes in their damned devil formation have to fly over, and turn all, all my cows!—into nothing, into— pieces, corpses! And the combine not moved in time. I want, I demand to know what is left in this kibbutz besides the dirt we're hiding in! Army experts. I had to shoot two cows. My Yaffa and my Shikmona. I can't stop hearing their groans."

His hard face melts, unable to reshape itself. At that moment, on the wireless, a phone rings—

—To test the rocket when it was finished, your father and I went to Rambam Hill. We walked hunched over to avoid snipers. How high was the hill?

Since you ask, it happens I have a story that will tell you exactly. Once I was walking in the Old City of Jerusalem, and a woman, a Christian, stopped me. She asked me where Calvary was.

"*What's that?*" *I asked.*

"*The most important mountain in Jerusalem.*" *Well, I tried every mountain I could think of—Scopus and so on—they weren't what she was looking for. Many years later, I visited the Basilica of the Holy Sepulcher and I saw their Mount Calvary behind a glass screen. A rocky mound, right? You've seen it? That's the height of Rambam Hill.*

A phone rings!—then the bunker walls dissolve like mist. The farmer and the officer find themselves looking straight into the lightless earth. In that subterranean darkness, eyes appear. Blinking, weaving to and fro. Hundreds of eyes, in hundreds of unseen heads. In blackness. In an endless chamber.

I remember, Itzik dug a launching pit very professionally, he kept repositioning the rocket until it rested at a good forty-five-degree slant. We lit the fuse. This was rocketry pared down to its basic principles: a container propelled by the force of escaping gases. That is a rocket. The rest is extra. We lit the fuse.

In blackness. In an endless chamber vast enough to contain a tempest: brown clouds raining dirt and broken boards, flying tin sheets and ghostly chicken feathers, whole fields raised in the air, and someone dangling by his foot from the top of a watchtower ladder for five days and nights, and across the vacant ground someone crawling, Do You Read, someone's hand cradling her pink bowel in the sand, Come In Do You Read, no help forever and sucked pebbles, taint in the wind, stretchers ammunition thirst grief thunder and are you sure there is not a single cigarette left not even a butt, not for me, for one of the wounded, he

has to have it? The horizon is on fire. The sky is a brown pit. Six
hundred shells per hour whizzing, hitting, blasting, Chaos itself—

—*. . . lit the fuse. The rocket flew! It flew about ten meters.*
—*Wait. That was the rocket?*
—*Which your father built.*
—*The one made from a water pipe? That was Israel's first rocket?*
—*Sharona. To you, and to me now, this dinky little rocket looks naive and
silly and like nothing at all. But you must remember that when this happened,
rockets were relatively new. The Germans had bombed London only five years
before. This was the coming technology, and we were doing it! We did it! It was
like, sort of, a miniature Manhattan Project. True, we didn't have the time or
resources to develop anything new. We barely had the time to copy what already
existed to save ourselves. But that rocket, I can't tell you how exciting it was!
Because it meant we had a future.*

—Chaos itself! Is contained here. Is bound by magic into a three-
act play, here, at the Dais Theater in Tel Aviv.

On the stage, which resembles an ordinary bunker, an actor,
playing an officer as worried as his audience, hangs up the wireless phone that tells him his position is surrounded, props himself with hands flat on the table, and says, as if reading off their
backs, "You ask what's left. The dirt we're sitting in. Yes, we are
crazy enough to love it and we have it to defend. Seven, ten days
from now, the arms shipment may come. We'll hold out, others
are depending on us. We will hold out, my friend. Because it is
not the tank that will win this war. It is the man who will win."

(All the actors' lines in this chapter are my loose adaptation of an
old Dais Theater script crossed with a book in which soldiers
talked about the Six Day War, private conversations, graffiti. I

too am an actress, a more or less secret impostor. I do art, not science, and this is my Telling.)

—I myself invented a device to measure rate of fire for Sten submachine guns. These guns were made in secret workshops, and the problem was, to standardize them.

— What does "rate of fire" mean?

—Just rate of fire.

—I don't know what that is.

—Right, sorry, your Hebrew is so good, I keep forgetting. You didn't do military service. And you really don't know what rate of fire is?

—I don't know a thing about weapons. I've never handled one.

—Really . . . Well, rate of fire, it's like the heartbeat of a gun. It's standard for a given automatic weapon, say, ten rounds per second. This is so you can teach soldiers to use machine guns—how much time between shots, how many shots, strategy, right? In the Gray House, we had a gramophone. I removed its playing arm, cut out a cardboard disk, put it on the turntable, and as it turned I would shoot one round at it. The bullets left an arc of holes in the cardboard. From the angles between the holes, and knowing the speed of rotation—78 rpm?—we could calculate the rate of fire. I used this device to test a gun for a Mr. Uziel Gal. . . .

—The Uzi?

—The same.

—Wow.

—This month, when I was in the cardiac unit and I lay there listening to the electronic monitors, I was thinking that my device was more accurate. Those monitors go by statistical updates. They don't clock every shot—that is to say, every beat.

—I want to visit the Rise. Misha, can you think of any reason why my father never told me this story?

—Forgive me, but I can just imagine what they'll say if an American girl

shows up, no security clearance, asking for a tour of the Rise. Let me give you some advice. These anecdotes are good for an evening. Granny's tales, bubbemeises. They're not immortal. Why don't you write about love? Isn't love at minimum sixty, seventy percent of poetry? That's a poet's subject, right? Love!

It's October 2001. Today my notebooks, this new one and the old Israeli pad, look like nothing but a collection of ink marks. I shove them aside. Leaving the house, I walk up the drive, sniffing cool air tinged with a neighbor's bonfire. I pass a meadow where coreopsis and daisies flowered all summer, and in it stands a bare walnut tree with a single nut, on a twig. The nut is lime-sized, lime-green, black-flecked, an archive of walnut genetics. About to fall on the drive and be crushed. Like the testicles of my poor dog who'll be neutered this week. Looking farther afield in the tall grass, I notice other leafless walnuts on which these green balls hang festively, a rehearsal for Christmas.

A thought stirs. I hasten back to the house, run into the study, and bend over the notepad, turning pages. Why did Misha compare Rambam Hill to Calvary, of all the unlikely things? I'd thought it had been the sort of puzzle an engineer would remember solving. But now I feel like I'm inhaling something very cold through open lips; and with an awful thrill of sadness, I read again the part where he teased me about love.

I never saw it before. I never saw it before. Misha was telling me the story of his youth. It's not what he meant to tell, but what slipped out of him in hints and metaphors. I think he must have avoided himself during that war so thoroughly, it has taken me all these years to read between the lines. *He detached the music-playing arm from the gramophone. . . .* On that long-ago evening, before leaving the-dinner table, Misha gave me a last, rather fumbling bit of advice. He didn't want, he said, for me to write "like we were heroes." This was not modesty. Close to death, he wished himself, not a false image, to be remembered. He hoped I would write something in which he'd recognize his springtime. What's that line of Alterman's? *Electric spring, sorrowful and bright.*

Misha, since we talked, you have left this earth, and the strip of it for which you strove with your wits, a little physics, and at minimum seventy percent love. I wish that as you lay in the cardiac unit, you hadn't had to compare your heart to a Sten submachine gun.

Part Three: To Hear the Truth of a Life

Nine

MAYBE THE "TALKS" I couldn't pry out of Uri Gil were like this one. I was a high school senior. My father and I would go to Bartlett's Sundae Shoppe, in the winter, when we were valued customers and could spend half an afternoon at a beige Formica table.

Scanning the laminated menu, my father, with his sallow, winter-worn complexion, made Bartlett's look like a place for children, or people whose lives were smooth as the Vanilla Creme Float. Somehow, Bartlett's also exaggerated his accent. He sounded like Bela Lugosi.

"Sviss cow," he told the waitress.

"A Swiss almond fudge," I translated simultaneously while my father went wide-eyed.

"Yah?! Is the one?"

"Yes, Daddy. And a chocolate fudge with butterscotch and no nuts."

"No nuts?" he gasped. "No nuts?" Nobody got his jokes. The waitress, like a pretty nurse, looked alertly from the weirdo to the junior weirdette. "But the good stuff, *nu*, the gooey—" He made taffy-pulling motions with his arms, and his nostrils flared. I wished he would at least remove his shabby woolen scarf.

"Marshmallow?" guessed the waitress. My father was ecstatic to know there would be marshmallow. While we waited for our sundaes, he unclipped a pen from his shirt pocket and tried to pull a napkin from the metal dispenser's slot.

"*Tchort!* Lousy design." He plucked out waxy shreds and pulled again with surgical delicacy. A meager napkin was extracted whole, and he began to draw on it, continuing our conversation from the car.

"Okay. Here is your physical body." A square-shouldered stick figure appeared on the napkin. "Around him, we are finding a mist composed of skin flakes, salt crystals, water." The figure acquired a body-hugging outline. I snuggled up closer to the table. "Around this mist there exists another body, known as the astral body, which extends to twenty-four inches or so. This, Pookli, is the so-called aura that psychics are seeing." Now the stick man, tightly outlined, seemed to rest in a sarcophagus decorated with radial stripes, like inlaid lapis.

I asked my father how he knew that the astral body existed. He glanced up without removing his pen from the napkin.

"What we are calling bodies, Pookli, is just a vay of talking about how we get around in other realities. This body"—he patted his broad chest—"is useful for getting around in one reality only, the physical. Period. But you know that you are existing in other realities, beyond the physical, hey?"

What gave him the power to ask a question that unfolded into a stairway falling infinitely below my feet, and rising above my amazed head higher than the meaning of height? If I hadn't already loved him for being my father, I would have loved him for that wonder. What could I answer, except that I knew I had always existed in many realities?

. . .

Two more shells were drawn around our stick man, as if, wrapped in gauze and interred in his sarcophagus, he now reposed in a boat. I sipped the foul-tasting ice water in my glass.

"Most people are active in only two realities," my father expounded in his casual way, "the physical and the astral. These people live for material things and for emotions, like your mummy. When we lose our physical bodies—what is called 'dying'—most people go to the astral and hang around there, becoming ghosts, wailing and clanking chains and so on. Is behavior of stupid uncreative types who have nothing better to do. But beyond the astral is a higher level called the mental. When you are in your mental body, Pookli, you have only to imagine a thing, and it appears. Like magic!"

I wondered what things he had caused to appear, for I knew he spoke from experience. It was one of the reasons why I was a disappointment to him. I had gotten myself accepted at Harvard, Yale, and Princeton, but my father's only comment on this, my life's work, was "Expensive." Which was guiltily undeniable. I knew he thought my straight As were boring; any plodder could get high marks from another plodder with credentials. Had I built anything? Had I laid pipe in the Negev? As for looks, he'd always refrained from suggesting that I might marry someday, probably out of tact, which also kept him from saying how plain I was, compared to someone like our waitress. Now if only I'd shown some native psychic talent, like his for traveling out-of-body, or a saving touch of telepathy. He'd insisted that I learn to meditate; it was during meditation that he zipped around in the other dimensions. But all I did was nap. In time, he promised, my nervous system would evolve to a higher level. Meanwhile, I cost too much.

. . .

"Finally"—he cleared his throat—"we are—creative people are—sometimes experiencing, for very brief periods of time, our existence in the causal reality, where you are in tune with the cosmos at the highest level." His tone became mild and sweet. "That is where poetry is coming from." He turned the napkin so that I could study the layers onioned around the little man.

"Poetry comes from emotions, Daddy."

He beamed, and almost, but not quite, reached to pinch my nose.

"Love, Pookli; love is the only emotion in the higher realities. Love is the material from vich you are making poems."

"But Daddy, love isn't just one emotion, it's complicated." I was sure that future experience would bear me out. My father's beaming bugged me, there was something askew in it. I did not venture to question his mental health. I habitually carried his cosmology around in my head next to all the schoolbook science I'd stored there, in a separate room, making an exception for him, not using my judgment. But this one point rankled—why, why couldn't he allow my feelings in his higher realities?

"Love," he said tenderly, "is not an emotion at all, properly speaking. It is a blue energy."

I would never be lovable to him.

"One Swiss almond fudge, one chocolate fudge?"

Scalloped paper mats were laid, and nicked flatware. The performance with which my father greeted his sundae was in jarring contrast to his low opinion of physical reality. He oohed embarrassingly. He shrank back from the goblet brimming with white goo, on its icy pedestal. He toyed with his spoon, measuring it in his fingers. Then he leaned over to whisper, "I am ashamed. I could eat two!"

Ten

WALKING THE DOGS, I noticed a large catnip leaf like a green heart trimmed in white lace. I reached down to feel it, and it crunched. First frost. With these winter fingertips, I pick up my Passover photograph, and the sight of Aia makes me feel old. How sweet she is, how pretty! Her slender calves levitate off her high white pumps. The women beside her look stodgy. She's clutching a big tote bag between the folds of her shawl, bringing wine perhaps, comradely and responsible, but so sweet! That's not the quality I remember. Though her presence is commanding—she shines, shines like a young tree hugging the breeze, her fluffy dark hair tossed over her shoulder! Did I say "sweetness"? I meant joy.

"It's open—come in, sit!" came a faint shout when I pressed Aia's doorbell. I walked into her salon. Greenish light filtered down from a plant-choked window into a room as bewildering as a Persian garden. On one wall hung a collection of antique ouds and melon-bellied mandolins, their tuning pegs carved with animal snouts. On a shelf, a row of Chinese urns enameled with splashy peonies. Across the cloud-pinks and teals of a Kirman carpet, like two islands sat a pair of capacious armchairs draped in Indian mirror tapestries, and between them nestled a Bedouin tray table of chased brass, its black legs inlaid with florets of mother-of-pearl. Cautiously sitting, I lowered my dusty handbag to the carpet. I thought that if this room were tumbling through outer space, without gravity, any of its sides could change places with the others without disturbing the overall effect.

Stifling an urge to inspect myself in the mirror cloth, I thought I felt a light touch, a caress . . . on my wrists sunken in lace, pearl buttons, rhinestones; I was fingering mysteries in the quilted tiers of her button box, I remembered. Love had mingled with its perfume. I had failed to remember for how long? How long since she had died? Ten years? My little Russian grandmother, Rose Rosenfield, who told me bedtime stories with happy endings, patting my hair. I used to think her sewing machine was called Singer because she sang when she sewed. How could I have forgotten? Grandma's button box. I loved my grandmother, but had lost the remembrance of it. Aia's salon had a scent; I sniffed it. Chicory, could be.

Aia made her entrance. She stopped in front of me, lifted my hands from my lap, and kissed my cheek.

"Just like him," she said, in a voice too large for her body. "Like two drops of water. Tell me how you take your coffee."

She wore a plum silk robe with an unraveling silver dragon on the back, gold bangles stacked to her bony elbows, and a big topaz ring. She went swishing in and out of the hallway, carrying drinks and sweets. An old hippie might dress like this, but Aia had an air, a marked air, of command. Was it her face? Square brow, rugged nose, thin painted lips, framed by hanks of dyed black hair. Her expression was brilliant and stern. After serving me, she curled into the other armchair, grasping her ankles and pulling them to the side, her formidable face nestled against the cushions.

"Sara told me about Sonia," she said. "You should know that, from now on, I will have no further relations with Sonia."

For various reasons, including a pearl necklace, I'd hated my father's wife, and I'd always been scared of her, but secretly. That Sara Gil had guessed correctly, that Aia would change her social life because of it, gave me a shock. I felt ashamed and started to stammer.

"No, it is necessary, it is necessary to care!" she overrode, brows meeting in a ledge. "To be involved! I believe in involvement. I am a kibbutznik! I do not believe in the competitive ethic, in jealousy, in possessiveness!" She watched me choke down a cookie. "I should have involved myself years ago. My husband and I visited your father in his home in Belleville."

"Belmont," I croaked.

"We went with him to an international fair, do you remember? Because you were with us."

"I—!"

"You went with us. We picked you up at your mother's. You were very excited to see your father. You jumped." Aia smiled. "I blame myself. I saw what they were doing, Itzik and his wife—ach, I should have intervened."

"But that was his private family matter—"

"No!" Aia swung her legs down and sat upright, massaging her bangles. "I believe in involvement! I should have said, 'Think of what is happening to the little girl! She will grow up without the circles of friends that expand, that keep you from feeling alone in the world!'"

For the second time in this room, I recognized something buried within myself, forgotten, or never known. Aia had named the way I felt. Which, until this moment, I had mistaken for the way it felt to live.

"Aia," I asked, "do you think he loved me?"

"What do you mean? You were his."

"What does that mean? Besides biologically?"

"He had his house, his wife. He had his work, which for Itzik was probably the biggest thing, his work. He had a child, you. You were his."

"I can't feel what that means, being 'his.'"

She gave me a wait-and-see smile, and I had a sense that life had changed, and that I should too. Pursue a new tack. But that would mean abandoning my plan, my plan for the Hollow Charge.

I had come to Aia full of questions about Hemmed's most beloved invention, the Hollow Charge. I had heard about it from Shai, a prominent physicist, at one of Sara Gil's dinners, and no sooner had he explained what it looked like, a bowl-shaped shell of spun brass, than I was hooked. There are images like master keys; once you get them, you have the way into a poem. You cannot know them in advance; they strike you, that's all, as the Hollow Charge struck me. I knew brass from my father's laboratory. I'd seen it turned on his lathe, and had played with the crunchy golden shavings in the lathe's disposal tray. My father, Hemmed, and I had all touched that brightness; maybe it stood for what connected us. I wanted, if not to see the actual weapon, to reconstruct it from the memories of those who had. I wanted Aia to describe the Hollow Charge in detail, so I could feel its heft, trace the score marks along its rim, know how TNT coated it.

With a firm image of the weapon, I would attempt a poem about Hemmed and my father. I would recover the vision that had eluded me for ten years, ten sterile, forgettable years.

I unclipped my pen and asked Aia about her work on the Hollow Charge; what it was like.

"What it was 'like'?" Aia echoed. "Let me think." She frowned and lifted her hand; the topaz emitted a juicy flash. "It was hard work. It was ideology. People went into Hemmed for ideological reasons."

"Not survival reasons?"

"That's American thinking. There is no such thing as simply survival. The work—yes, it was hot, the conditions were poor and dangerous, but it wasn't just necessary—it made you happy, like the pioneers who broke rocks for the roads and felt they were building the Land. I came from the kibbutz!" she exclaimed. "I came from Ein Harod!" She sought my eyes, and I hid frustration. Get her talking freely, then she'd spill the details.

"Your work today. Did Hemmed influence it at all?"

"I studied physics in order to learn the nature of music. You should write this down," she urged. "What *is* music?"

I wrote it down.

"Musical sounds do not exist in nature. No such thing! We invent the raw material of music, the notes and the scales. I study sounds to find out what gives them their meaning."

Aia looked pointedly at my notepad; I stopped tapping my pen on my lip, and wrote.

"Does the meaning come from the universal responses of our nervous systems? Or is the meaning given to the sounds by our cultures?" Rhetorically, she demanded, "What *is* a musical sound? What *is* a tone? What *is* a note?"

And she lectured for half an hour.

Every moment robbed me of the Hollow Charge. My notes shrank to listless jots, and I was anxious to the point of tears. To distract myself I wondered about Aia. It would be good to be strong like her. She came from Ein Harod, the legendary mother of the kibbutzim; it was the first kibbutz to raise children in a communal children's house, where they learned from the first nannies (Rivka, Frumka, Natalka, still remembered) that life was cooperative, that no one should be alone in the world.

"Excuse me." I broke into Aia's lecture. She had digressed to the topic of recording people's brain waves in response to birdcalls. "How exactly does birdsong convey emotion? I mean, what is the mechanism?"

"By the tension and relaxation of the sound," she said with teacherly approval. "Birdsong tells us a sound's universal meaning, the one rooted in nature. We know what birds mean when they sing because we know their behavior—they are driving off a threat or courting a mate—and birds do not lie."

"Birds don't lie."

"No."

"Did human beings once sing instead of talking, but started talking when somebody told a lie?"

A phone rang in another room. Aia ignored it.

"What we surmise," she replied, "is that music and speech became gradually . . . different." She checked her watch. "I hope this has been helpful to you?"

"Very helpful, I'm so grateful," I lied. "Aia, if in the time I'm here—that is, if it's not an imposition—"

"We will continue," she said, rising, cordial.

"Thanks, but I was wondering, if you happened to remember any other stories about the war, would you mind very much, if you felt comfortable—" My time was up. "I'm trying to understand my father better," I said in confusion.

Aia billowed back down onto the edge of her chair.

"It would seem that I have wasted your time."

"Oh no! Not at all."

The phone rang again and she ignored it, looking up at me, considering, her hard wrists hanging over her silken knees.

"It was a long time ago," she warned, waving me back to my seat. "I was one of the defenders of the Jewish Quarter in the Old City."

"You were!"

"I was. Before I joined Hemmed."

"Oh."

How do we hear the truth of a life? With our ears, jaded by lies? In memory, which is unreliable? How well do I hear Aia? I was silent, out of respect for what I couldn't imagine enduring. Even that's a lie. I imagine it all—how can I help it? The ancient Jewish Quarter sealed inside the Old City walls, denied food, fuel, and medicine for the final six months of British rule, then conquered by the Arab Legion: the siege-within-the-siege. I can't help hearing shrill calls of children, famished, scavenging the cobbles for empty cigarette tins to make into bombs. The wireless dispatches hourly imploring a command with its hands tied. The crunch of stone courts that crumble with the shaking of the earth. In Glory of Israel synagogue, Aia hides in the women's gallery, hurridly circles the gilded dome, her palm to the drumming wall, two dented bullets left in her pocket. Chandeliers rain crystal. Benches crash on the ground floor lost in smoke, fires ransacking the air, and grappling forms that she barely identifies. As she coughs like a jackal, crouches, and reloads, is there a song in the golden dome of risen prayers? A universal sound for her? On the wall beside her glitters the Tree of Life, mosaics of pyrite, olivine, cinnabar. Slowly exploding.

Aia's braceleted arms are in her lap, her face thoughtful.

"I commanded a sector of the rooftops. They were mostly— in the Jewish Quarter—old people and rabbinical students, help- less people. They had talismans against fear: 'Say a psalm and the bullet will miss you and hit the wall.'" She looked ironic. "I had no fear, I was fearless. Only later I had dreams. I commanded a squad of Kurds. Wild men! The sheepskins they wore, the way they cursed each other. I acted utterly hard and ruled them. There was no other way for a girl to take charge. But all that changed one night, after the death of a man who had been my friend. It was the first time the war really hit me; not before, despite the siege conditions, the shooting and the starvation. One of the Kurds walked up to me. He saw that I had tears on my face."

Burdened with its topaz, her thin hand touched her cheek.

"Discipline went out the window, and I couldn't rule them afterwards. But I never felt the need for . . . The Kurd, when his girlfriend was killed, he shot an old Arab woman for revenge. I was shocked. Not only the brutality. A life cannot pay for a life! Life is not a possession."

Her tone was candid and gentle. I wrote it all down on the page where I'd thought to have possessed the complete descrip- tion of the Hollow Charge, my golden image—oh.

I see. I think so.

Wasn't it an idol, a false idol? I had treated vision like a possession, something all for myself. I might as well have tried to possess the scent of chicory, the verdant light, and time passing . . . as I recall, I left unsatisfied. Maybe it was just my American thinking.

Eleven

IN THIS STACK of blue aerograms lies the story of how I came to be. Once, I tried reading them, and by ill luck ran across the phrase "bar-mitzvah-age son." My fingers felt sliced up, so I stopped. That was a few years ago, and as Tom says, time's a-wasting! Without reading these letters, I'll never understand the world into which I popped, out of the void, into a northern Massachusetts blizzard, into the arms of a callow immigrant who was becoming an American success, Invention-a-Minute Ben.

These are the letters my father sent to Dan Yaari, his friend from the "Negev Beasts." The Beasts were a commando battalion famous for their motorbikes, Aussie slouch hats, and the rest of their uniform, consisting of shorts and bushy beards. They would roar into a kibbutz, hog the communal shower, raise the roof of the dining hall with their gnashing and guzzling, remount their bikes, and zigzag off into the dunes, leaving behind military intelligence and other kinds of gossip. My father wasn't a Beast (not his style), but he and Dan were pals. When I met Yaari, he was a laconic manufacturer who looked with habitual tenderness upon his wife, who handed me these letters tearfully, cheerfully, and without ceasing to chatter for several hours. For years afterwards, I'd get overseas calls from Mrs. Yaari.

"When are you coming? You promised! Last time we spoke you promised. You must take a vacation. You must visit. Nonsense. I'll pay! We were his *hevrei!*" Mrs. Yaari would cry.

"The history of every country begins in the heart of a man or a woman," wrote an American author. Israel began in the hearts of the hevrei. Without this slang term, Israel's beginnings are no more comprehensible than those of the United States without the word "liberty." There is no translation. Here are nine English equivalents and the reasons they don't work:

1) Group: too generic.
2) Gang: too gangsterish.
3) Clan: too feudal.
4) Tribe: too archaic.
5) Band: too nomadic.
6) Society: too vague.
7) Collective: true only of the kibbutz, the logical extreme of hevrei-ness.
8) Pack: too zoological.
9) Y'all: not a noun.

The way an American veteran feels about his platoon buddies: extended to include both sexes, every aspect of life, and one's entire lifetime. That might come close. The people you played with; went to school with; served in the army with; married among; had kids at the same time as; went into business or to the university with; criticized; consoled; dealt with and felt with, and could not imagine yourself, the insides of your head, without. Hevrei.

.   .   .

A singular noun that takes plural verb forms.

Here is the first brittle aerogram, crammed up to its shiny adhesive margins with his round Hebrew script. He wrote from Leominster, a village in northern Massachusetts, where he'd been hired at a Jewish-owned factory. My father's friends had gotten him this job, because Israel, a socialist statelet without a dime to its name, in the worst depression of its history, was no place for hi-tech entrepreneurs. And he didn't want to make weapons anymore.

> I Bentov
> 50 West St.
> Leominster, Mass.
>
> 10.30.54
> Shalom Hevrei!
> Nu, finally I can settle down and write. I'm here in a pretty little town, 24,000 inhabitants distributed over a vast territory. The houses are colorful and made of wood. The churches are old and made of brick. Everything here is designed for the convenience of people, unbelievable. For instance: you buy yellow cheese, open the package, and it's sliced already. Bread too. It's a beautiful job of slicing and saves much time. . . . You buy hot coffee/tea, put it in your pocket and go. All fluids are obtainable in sealed containers to take with you and not worry about spillage. Generally, there's a huge waste of paper. You can't buy a handkerchief, instead they make handkerchieves out of very soft paper that you use once and throw away. There are towels made out of paper in public places. . . .
> I'm to develop a machine for the factory, they produce plastic buttons and crap like that. My job is to build something that will make 110,000 plastic pearls (no joke—a hundred ten thousand pearls per hour).
> Afterwards, to build a machine to string them together. Nu, a living . . .
> —Itzik

The next letter features his first American word. It follows
his discovery that the workweek in the U.S. was not Israel's biblical six days. Like most people who have to learn a series of
adopted tongues, my father cut corners, and never wrote the
English lower-case. His American words jut from their Hebrew
background like an aerial view of the tops of skyscrapers.

> *11.21.54*
> *Shalom Hevrei!*
> *. . . in another 2–3 weeks I hope to put my machine into operation and
> it will introduce (if it works) a little revolution in the manufacture of some
> of their products. I hope also to improve my salary, since my boss is fairly
> rotten in these matters and exploits my being Israeli and pays me less. . . .
> On the other hand, I have the use of the company car to drive into Boston
> for my WEEKENDS. . . . I have met hevrei from the Hemmed studying
> at the universities. . . .*
>
> *This week, I've decided to buy a radio. I don't hear the news or read
> papers. I heard the Chinese invaded Formosa—is it true? Here, news-
> papers generally report a theft or a murder, the movie schedule, and the
> results of football or baseball games. . . .*
>
> *I can get a decent radio for a week's salary. There are special stations
> that play classical music, and I can't go without it anymore. All the rest is
> such crap you can't listen to it.*

It was tough on Itzik to meet hevrei who'd been sent abroad
to the best schools, by the Israeli government, with the aim of
their returning home and creating nuclear resources. While
Itzik—did what? Made plastic pearl necklaces. Where? In a town
forgotten by God. He had reason to be irked. But to ignore
American music! I have a bad feeling about my father's starting
off this way.

Alone in his room, reading *Scientific American* by the light of a secondhand lamp. Outside the northern sky is black, and the houses are built on frozen granite. Yet everything he needs is in this hard subzero air, potions for the homesick and the lonesome, swirling in his drafty garret. There's Sinatra for cool nerve, there's Como's croon for relaxation and Satchmo's brass for stimulation, and the Grand Ol' Opry for salt-spangled balladry and Tito Puente with his mambo in the Catskills. Daddy! At your window, by the icicle thicket that hangs from the eaves, floats a sharkskin suit, cut from starlight and just your size. If only you would look. Stop chasing the BSO around the radio dial, give the kapellmeisters a rest, let go of your imported European consolations. Europe! Just how many homes are you homesick for?

Well, here's a letter to his big sister, his sole surviving family member. She went to Palestine before 1933. When I met her, in 1968, she was a heavy, ground-down creature living in a farming cooperative with her family. I didn't take to her, poor soul, because she whined constantly about the nice life she'd lost in Czechoslovakia, where she'd played competitive tennis and had maids. In her bedroom I glimpsed oval-framed portraits, and was awed to think they might be my grandparents, the dead ones; but when I asked, she jiggled all over with a jeering cackle that seemed aimed at someone who wasn't in the room. After I returned to the States we lost contact. During my Hemmed summer, I did call and say hello. Not to my aunt, by then hospitalized, but to her eldest daughter. We had a friendly chat, and I asked if our grandparents had left any letters. My cousin gave a shivery laugh. There were letters, in Hungarian, that my aunt used to lip-read and cry over.

"They were frightening to behold! They were sent from the camp and the Nazis cut holes in them—they looked like lace!"

"I can have them translated," I offered eagerly.

"I begged her, Sharona, I begged her. She burned them."

My cousin did oblige by sending me the following letter that my father had written in Leominster.

> *10.15.54*
>
> *Shalom D——!*
>
> *So there won't be a great outcry I'm replying immediately. . . .*
>
> *It's a shame you hired a lousy builder. I don't understand what you mean by the floor "sinking." It sounds like a foundation problem. . . . Supply details. If I were there I could look for myself so you must supply details. You say judging from the rainfall it's a drought year, well I hope you haven't sunk any more into some agricultural outlay or other. I understand the chicken run is coming along. Nu, nu, we'll see. . . . I'm sending you some gloves made out of rubber, for washing dishes or outside work. . . .*
>
> *Here the temperature went down to −25 C, it's strange to go about in such cold. The trees are frozen and their twigs look like whiskers, beautiful. Today I took the car and drove to a village that's not on the maps, about 160 km. to visit an Israeli couple, although there are no leaves on the trees. The mountains are wonderful and the wooden houses are colorful and very pretty, and I came back half an hour ago and I don't understand why I have such a headache.*

Driving in dead Canadian cold, hundreds of miles over mountain roads, to see hevrei. His car in winter was as personal as a Yukon miner's shack. To the dash he'd affixed plastic compartments for Kleenex, Sucrets, and Palgin, an Israeli aspirin, since American aspirin did not kill his pain. On his hands he wore, first, thin woolen gloves, then shearling gloves, cut so that each finger, in wool underwear, poked from a sheath of yellow fleece. With his neck wound in a tattered scarf, stubble standing out on his color-drained face, my father in wintertime looked positively orphaned.

Returning to the aerograms, I find Itzik ever more anxious to hear from his hevrei.

12.27.54

Shalom Hevrei!

[Dan,] you rotter, you have not a drop of conscience, why don't you sit down, rotter that you are, and write without a conscience. . . .

The first thing is to drink a little. From this they become happier and then all goes more smoothly. At first I couldn't believe it, until the chief engineer, a Jew who understands these matters, shut me up and poured me various liquids and it truly helped. . . . At the XMAS PARTY it is customary to kiss in honor of the holiday . . . nu, I too did not withdraw my hand from the plate, and kisses flew. . . .

The chief engineer might not have been Jewish; in Israel "a Jew" means "a dude." Otherwise this letter was scandalous. In those days the hevrei drank beer to cool off, and wine on festive occasions. They had no concept of social mixology. Sex they smiled upon, but in the dark. Drunken kissing in full view? *Tfoo!*

2.6.55

Shalom Hevrei!

If it's so hard to write a letter together, why don't you each write separately and put them in one envelope and send them? . . .

Outside the factory there's a large parking lot which is full up with the workers' cars. And what cars! It's a craze with them, they compete over who buys the most expensive car. . . . I have to adjust my ideas of travel and dis-

*tance. A half-hour's ride here, a trip of 40 km., is not called a trip. A jour-
ney like the one from my place to Boston, about 85 km., is called "to go into."
Moderate speed isn't allowed. For instance when I took the boss's car to New
York, I did the 400-km. drive in one breath, 4 1/2 hours, and they said I
drove slowly. But I'm beginning to get used to this too.*

This frigid planet of whizzing speeds, dizzying drinks, sliced
foods, public lips, plastic jewels by the ton, and mountain-
descending rivers of giant cars. In Israel, you could stand in many
spots and look into another country, which, it's true, wanted to
kill you, but within your bearings. Itzik was marooned on Mars.

5.7.55
Shalom Hevrei!
*. . . how is it possible for you not to write for so long, you're almost my
only tie to the Land, so you are under a national obligation to write to me
now and then (not once in half a year, curse you!).*
A) Work:
*. . . the boss is a genuine rotter and I'm leaving the first chance I get.
I'm building a machine to replace INJECTION MOLDING of some
products. . . . I'm independent, there's nobody over my head except the
factory owner, but he shortens my life enough!*
B) Fun:
*So-so. I've somehow become mixed up with a group from the Inter-
national Student Center where I spend WEEKENDS in Boston. . . .
We travel near and far, near meaning up to 150 km. and far, 500. . . .
I've bought a PLYMOUTH 1952. . . . there's a Danish girl, an English-
woman, an American, a Chinese, a Turk, and I represent Israel. . . .*
C) Girls:
*So-so. The American girls are worthless. Spoiled, dumb, narrow-
minded. Boring. The Jewish girls among them are the worst. Generally,*

I'm becoming more and more anti-Semitic and I'm surprised there isn't more anti-Semitism here. . . .

I read this, and I'm like, uh-*huh*. His character had begun to acquire protective camouflage: he wanted a bigger salary, better benefits, and more fun. And fewer Jews.

As for his ungallant remarks about my countrywomen, I can translate those readily enough. "Spoiled"—now, that might mean "too expensive to date on a green-card salary." For "stupid," how about "too smart to act brainy on a date"? "Narrow-minded," in 1955, in white-gloved, pearl-chokered Boston, home of the Boston ban? Ulp.

For the first time, Itzik's success with women, which had been perfect to the point of inconvenience, had gone sour. He no longer attracted fully-fascinated girls by the simple expedient of walking across their field of vision. No longer did human tigresses give chase. Holding a chilly cocktail glass, he would edge up to circles of partygoers wearing garments for which he had no names, all of them yacketing in American. Between their elbows, he would wait to overhear some topic that he partially under-stood: the Kennedys? He'd honk to clear his throat.

"Ehhh, Protestants and Cat'oleeks, excuse me, zey are all Christians, right? Ehhh, can you please to explain what is zeh difference?"

In Israel, any girl you met was like your cousin. If she was a kib-butznik, she was hevrei. If not, you still knew hevrei whom she knew, from her city (there were only three) or her college (there were two) or her army corps. You listened to the same radio broadcasts (there was one station) of the Israel Philharmonic or

the comic routines of Chizbatron. You hiked the same mountains, swam in the same lake (there was roughly one), and sang
the same songs (which were innumerable). What gave a girl mystery was the thought that she must, in her heart, have unshared
thoughts.

5.30.55
Shalom Hevrei!

*What's new? Heat! It's terribly hot. Disgustingly hot, hotter than at
home and damp something fearful. I sit and sweat all day, the room is a
steambath, you can't exist without a healthy fan. It's simply unbelievable. I
got back last night from New York, there people lose the wish to live. The
hazy heat and the soot that clogs the air, between them, in 1–2 hours, turn
your collar pitch-black from soot mixed with sweat, in short, filthy. Even a
sense of humor doesn't help. Where oh where is the pleasant, temperate
Israeli summer?*

He wants to go home. He wants to go home, but how? Then
Israel strikes oil (one well).

9.26.55
Shalom Hevrei!

*Since it's Yom Kippur and it behooves us to forgive and render mercy
to each and every one, I'm writing to you . . . Mazel tov! We struck
oil! . . . don't be a rotter, inquire around and let me know if there's any
demand for mechanical engineers or petroleum PRODUCT DESIGN
and maybe I can take part in it. . . . I would so much like to know . . . if
it's serious, I'm leaving here and the shitty pearls . . . and I'm coming
home.*

*I invested considerable money in my devices and the outlook isn't
good . . . actually, I'm BROKE and I can't even pay the dentist, may his*

name and memory be erased. . . . I'm very RESTLESS and can't concen-
trate. Working like a donkey. Too tired to sleep well, and too confused.

In the midst of this sad, begging, hopeful letter in Hebrew
script, there stands out a pure hard fragment of American poetry:

> *Product Design—*
> *Broke—*
> *Restless—*

When you deny the music of a new homeland, it will play
upon you in its own voice.

> *11.17.55*
> *Shalom Hevrei!*
> *I rejoiced to get your combined letter, and it came on the same day as a*
> *letter from S——, so I am now enjoying complete contentment. . . .*
> *I've decided to stay in the profession of plastics. It's fascinating, though I*
> *don't know what I'll do with it in the Land. Here they make everything out*
> *of plastic materials reinforced with fibers of glass. Heavy things. Small boats,*
> *auto bodies, bathtubs, in short everything and I'm very drawn to it. . . .*
> *I've also decided (Hear, oh hear!) to get married this year, that's that.*
> *Why? Because it's making me nervous already. Possibly it will make me*
> *more nervous afterwards, but what can you do.*

This letter gives me goose bumps because I see in it my own
conception, in my father's desire to banish nerves with the deliberate acquisition of that ambiguous asset, a wife.

Itzik met Miriam during a WEEKEND, at a party given by graduate students in a crowded Cambridge apartment. My mother

was in a corner, being startlingly conspicuous. She was five foot ten, with long, straight hair as glassily black as the belt around her eighteen-inch waist. Her eyes flashed; her voice was cultivated; her gestures imperious. But when she laughed, she was like a little girl hanging from a jungle gym and grinning upside down. Around her clustered her typical circle of admirers: fidgety intellectuals who lived on talk, parasites sniffing around for a naive hostess, and here and there, someone so modest, so devoted to her animal gaiety and rebellious mind that he was almost bound to be overlooked.

Plus the onetime fluke, my father.

Itzik was impressed. He missed discussing ideas, and Miriam had a bachelor's degree in philosophy on top of twelve years of Talmud school; he missed speaking Hebrew, and she spoke Hebrew, albeit a tad Talmudically. Within fifteen feet of her he knew that she had fire and magnetism enough to warm up his room in Leominster and reorient his north. He was thirty-two, she was twenty-five; they shared purity of heart, and about themselves had not a clue.

"He's a desert rat!" argued Miriam's roommates. "He's uncivilized!"

"He's a breath of fresh air," she laughed, but what she meant was the incident of her boatneck sweater. Nothing is quite like a boatneck sweater, the tension in those corners at the bare verge of the shoulders. One night when they were sitting together, elbows on the table, Itzik plucked her left sleeve between his fingers, pulling it back slightly so the fabric wrinkled, no longer stretched tight. Then he sighed and said, "Now I feel better."

Miriam immediately adjusted her other sleeve to make life gentle and comfortable for it. If he had empathy for a knit, how much more for the woman inside it? And unlike her tyrant father, Itzik never raised his voice.

In my mother's tale of the marriage proposal, there is, to my ear, already a fatal hint. They were watching a show in the Hayden Planetarium at the Museum of Science. Out of the dark, my father said shyly, "*Nu,* so, when are we getting married?"

"Huh? We're getting 'married'?" Miriam was nonplussed. He had asked her once before, but somehow, she'd thought he was joking.

"Aren't we getting married?"

"Okay, we'll get 'married,'" she replied. With the disbelief of a woman who doubts her father's love.

On their wedding day, my uncle Bert (ex-GI, BU law school) went hunting through the fourteen rooms of my grandparents' house. The rabbi was waiting. The parents and relatives were assembled and seated, even upon the velvet sofa's buttock-prodding buttons. My mother sat hiding her veiled face in her hands, in an upstairs bedroom. Finally, Bert strode into the attic and found my father deep in *The New York Times.*

"Is time to get married now?" Itzik asked.

"Geeez," apostrophized Bert. "Will you put down the paypah?!"

A few minutes later, my mother became the bride in the oil painting that used to hang over my grandmother's piano: her hair upswept, her face young and spacey, as she dragged the train of her gown, encrusted with seed pearls, down the stairs and into the TV den. My father, wearing a tie for the second time in his life (the first, passport photo), waited underneath a nuptial canopy's white plastic trellis and white plastic flowers. Furtively, Itzik was pulling at a plastic rambling rose.

Miriam stepped up to his side. Ooh, murmured aunts and cousins in approval of the bride's satin pooled on the carpet, and the shining column of her back. Unseen to them, Itzik began scraping the plastic flower with his thumbnail, absorbed, while my grandparents' ancient little rabbi (of whom it was always said, "Now, that's a rabbi!") quavered through the liturgy and came to a pause. And paused. With the point of her satin heel, Miriam jabbed Itzik's instep. The rest took place in whispers:

"Ai! What? What?"

"Not now!"

"Oh! Sorry! I've never seen such a polystyrene!"

After the ceremony and the feast, as Rose swept the honey cake crumbs from the linen, but before the newlyweds departed in his PLYMOUTH, they decided to open the wedding gifts and store them at Miriam's parents' house while on their honeymoon. The couple decamped to Bert's room, and heaped their wealth on the bed beneath the Red Sox pennants. First they opened the envelopes. Fives, tens, and a lone fifty-dollar check were extracted. Itzik, his tie unknotted, scoured the offending envelopes with a limber forefinger and expressed, to his unveiled bride, disgust. She was ashamed. Without raising his voice, he called her family "petit-bourgeois." The slashing retorts she would have mustered in any other dispute were laid by, for me to pick up a generation later:

Wasn't he a fresh bridegroom, with a lovely young wife?

Wasn't he a Holocaust orphan, welcomed into a thriving family who hoped to love him?

Wasn't he a kibbutznik who shunned money-grubbing materialism?

Wasn't he a Zionist who had turned his back on the customs of the ghetto, including dowries?

I keep returning to that sniffy "petit-bourgeois." It betrays him to have been, after all, Imre Tobias, a provincial Slovak whose values were European. He aspired to belong to the mon-eyed, cultured, assimilated gentry of the town, not the little Jew-ish shopkeepers. When I think of my grandparents sleeping five hours a night for the years it took to build their little bakery, rais-ing college graduates on a Depression diet of barley-and-bone soup, stout-hearted yeomanry of American business—what a yawning ignorance was his!

They left on their honeymoon; he wrote to the Yaaris.

1.6.56

Shalom Hevrei!

So, not in vain did I threaten you. It's true—I got made a groom of, no sweat. So I have a nice enough girl. Her name is Miriam. Brunette, tall, and as I said pretty good, at any rate above average.

We had a week's honeymoon and we traveled to Washington, the capi-tal, to do honor to Lincoln and all the hevrei who died in those good old days. We ran through all the museums. It was cold and dismal. And in the MOTEL the central heating went off and we froze all night ($-15°C$). By the way, MOTEL is an abbreviation of MOTOR HOTEL, which is like many small cabins beside which you park the car, so you don't have to drag your luggage from a distant spot.

We've moved into an apartment of two rooms and a kitchen. A pretty crappy apartment. But it's harder to get something better here, even for this crap we pay $18 a week.

The main thing—I'm learning PLASTIC TOOLING and REIN-FORCED PLASTICS, a field that's developing here at an awesome rate . . . there are thousands of things that need to be made lighter, quieter, color that's indelible, rust-free, and so on. All this is to be done with REIN-FORCED PLASTICS. Sinks, chairs, cups . . . [he writes in such excited haste, his script's illegible for some lines] *. . . light and cheap and easy to make. The steel that goes into everything is less manage-able and there's always the problem of thermal conduction. Here they just produce the product in prepared molds, already in the final shape of the product, and there's no need to work expensive steel, no need for steel at all. . . . I'm very excited about all this!*

Can you write me . . . what a ton of steel costs, and a ton of iron, in the Land? What's the hourly wage of a steelworker? From these parameters I'll be able more or less to calculate if the thing's worth it or not in the Land. So next time, write to me about it . . . yes?! And now how are you, and how are the chickens laying? I hope you won't have the same problems as my sister. And now, hevrei, shalom to you and lots of greetings to each and every one, and to S——, what's with him, has he gone away yet?

I rise from my desk. I take this last aerogram and go into the kitchen, where my husband is stirring a beef stew that I made earlier today. He's lifting the ladle to inspect the melting beef and carnelian carrots, and to inhale the flavors of bay leaf, anise, caraway, and peppercorns mingled in the vapors clouding his face.

"Sweetie, wanna hear something funny?" I ask him.

"Sure do." I read the letter aloud, translating. Tom lets out a whoop.

"I can't believe he would write that! 'Above average,' what a—I just can't believe it! What a character. You can't ever let your mom see that letter."

"You like my stew?"

"It's goo-ood." I hug him and go back to work.

. . .

He remembers a freezing motel. She remembers that he was too cheap to pay for the heated one across the road. Anyhow, what can be learned from an occasion like a honeymoon, which exists to defy repetition? Tom's and mine was a day long. We drove six miles to the nearest river town, slept holding hands in the Mill House, and took a ride on a historic canal boat. I dozed on my husband's shoulder, the covered boat was so warm, and the water rose, rocking, so gradually in the lock, while a youth in period dress, cheeks fuchsia from the heat of his derby hat, pretended to speak his mind about President Grant.

My father's next letter was written during the Suez campaign, but even into this serious business, he insinuated the name of his passion, his Shulamith, his north star.

> 4.11.56
> *Shalom Hevrei!*
> *We're glued to the radio and trying to wring news of the Land out of it. The rotters at the Consulate won't give any information. Poor Miriam has fallen ill from emotion. S—— wrote twice but hasn't shown up, I'm still waiting for him to visit. [Ora,] I assume [Dan] is not at home, so why don't you write to me fast and let me know how you are, I'm very worried. . . . Because of this war, plans are in flux, I'd thought of starting a new department that would expand the Sahan Company of Rehovot into REIN-FORCED PLASTICS. . . .*

I might enjoy this picture of my parents' vigil around the radio, hands clasped perhaps, as the announcer speechifies through their two rooms, if only I weren't sick of the Godot-like, the eternally anticipated S——, whoever he was. Judge for yourself whether or not I have reason to resent S— from day one. Here is a greeting card. The front shows a bedraggled cartoon couple holding a baby bottle and pacifier, and the caption "We've had it!" My father's script covers the inside of this announcement of my birth.

1.20.57

Shalom Hevrei!

Oy, what trouble it is to be a daddy! What's done can never be undone! If you only saw me now: Itzík Bentov, a settled man, a father in Israel, tending a screaming infant. I've had an average of four hours' sleep per night for a week and as a result we both have the flu. And the family descended on us. And we have no help. You can't get home help here because the factories pay more. People in our situation can't afford such luxuries. I haven't gone to work in two days, which is why I have some time now. At work, things have improved. It's very possible that we'll start producing REINFORCED PLASTICS. That's just what I've been waiting for and working toward. I see my Hebrew's degenerating. S— wrote me some postcards from New York, he's gone all around the world already. He was in Cuba, California, Mexico, and wherever you like. But he didn't visit us. Mainly, I hope you're feeling better than I am.

Who was more welcome in my father's life, eh? I wish, S—, that you had visited—I would have liked to spit up on your shoulder, S—! Nevertheless, when my birthday rolls around, my mother always tells me, "In the maternity ward I hadn't bathed yet, I must have looked and smelled terrible, he hugged me and kissed me, and for the first time I saw, really saw, something behind those eyes. Because his eyes always looked sort of— like there wasn't a way into him. But this time there was. And I'll never forget—he said, '*Ani m'od ohev otah.*'"

"*I love you very much.*"

Look at this snapshot dated 1958. This would be near MIT, in one of those factories whose brick walls advertised products in gigantic lettering: TOOL & DIE or PIPE FITTINGS. They said what they made and they made what they said. The Golden Age of Hi-Tech! The office walls are cream above the strip of wood trim, brown below, a perpetual reminder to drink coffee. I can smell that lead-based paint. I can hear that old radiator clank. Engineer One, a slouch, rests his feet on a chair, holding a meerschaum pipe. Engineer Two raises dark Latin eyebrows in grimacing interest. What sets my father apart, standing at the flip chart, isn't his absent tie so much as his stance. One hip leans toward a relaxed hand; the other hand, at breast height, displays a forefinger drawn apart from the rest, not quite beckoning. An attitude of classical statuary, known to Athens and to Rome, has been revived in my father's technical presentation. He's happy. When the camera caught him, he had been smiling for some time. The engineers are on the verge of exclaiming "Say—" and "Look, suppose—" and "Why the hell not?"

6.6.58
Shalom Hevrei!
Haven't heard from you in a long time . . . did I tell you about my new job? . . . I'm in CAMBRIDGE working for DEWEY & ALMY CHEMICALS, a non-Jewish firm. . . . I work in what's called PROCESS DEVELOPMENT—that is, once the labs come out with a product, I figure out ways to manufacture it commercially. . . . It's a great opportunity to learn about efficient systems of labor. . . . The company makes only trade products and not all sorts of dreck like buttons and pearls.

Passover, 1948

Itzik and another Hemmed member at the Weizmann Institute of Science

Itzik and Ziva, 1946

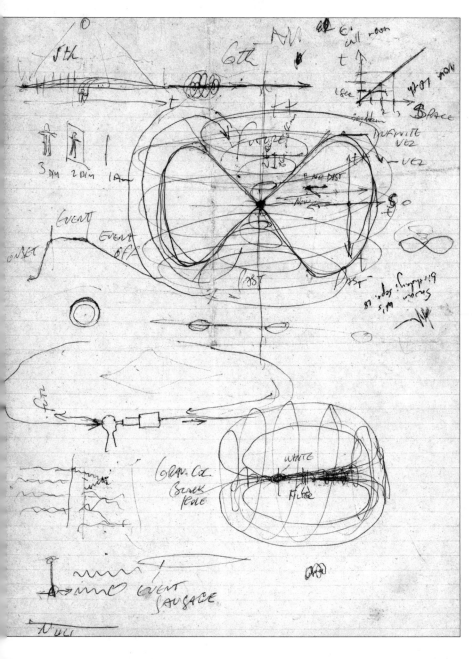

Drawings done by my father, while explaining the structure of space, time,
and the cosmos to me, one Saturday afternoon around 1974

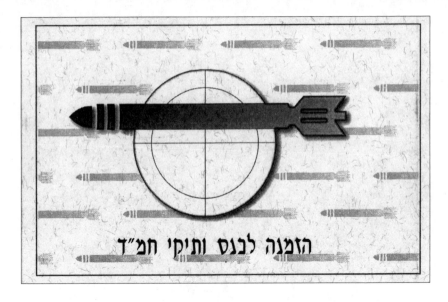

My invitation to the Hemmed reunion, showing the emblem that the Corps invented for itself during the '48 war

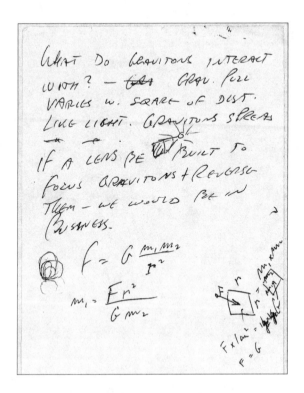

Notes for an antigravity device

Itzik in the mid-1950s

Invention-a-Minute Ben

A hand-drawn valentine from my father, around 1977

The Loretta

Imre Tobias and his family, around 1933

My father and myself, on an outing in 1960

I'm exposed here to the latest materials. . . . I also learn ECONOMICS—
that is, to compare various systems and see which is the best and the most
profitable, cash-wise. You have to calculate investments, loss, and so on.
This way I'm learning how to build a BUSSINESS. . . .

When I add up the photograph and the letter, the sum is
a successful pursuit of happiness. Itzik Bentov had become
Invention-a-Minute Ben.

NOVEMBER 17, 2001

I've avoided this next translation job for twenty-four hours.
Last night, as I was lying awake, I heard strange squeaks, like dis-
tressed rubber, coming from the hallway outside the bedroom. I
rose and felt my way along the hall until my toes touched fur, lost
it. I stood still, wrestling with the impression of a long tube of
superimposed dogs, a probabilistic topology, a topolo-doggie,
stretching toward me. . . . There he was—nearly kicked him.
Poor boy, having a bad dream. He lifted a forepaw hopefully, and
I rubbed his chest, deeply keeled, remembering the night we
drove home all the way from the Chicago seat of the German
Shepherd Rescue. It was icy and rainy, like tonight. As I rode in
the back seat, the dog's warm tongue kept swiping my face, to
taste who I was. He'd been left to starve. He'd weighed seventy
pounds, most of it seemingly in his trowel-sized ears; now he
weighed a good hundred. His paw slipped over my wrist, and
with a last pat I returned to bed. Tom, awakened, asked about
Mister Mix. Getting only mumbles from his wife, he went to
make herbal tea, waking Daisy, who started to pant in her rattled
cage. Black, chic Daisy, mistress of hullaballoo. Whining in husky,

speaking tones about her duty to follow Tom into the kitchen and defend him from peril. She takes everything to heart. When sufficiently worked up, she bites her cage until the welds part. I thought of letting her out, but at night she tends to spit up. I revolved this simple problem round and round as I drifted toward sleep. Mister Mix tromped in and collapsed, shaking my side of the bed, wafting his smell of salt popcorn and fresh dirt, and issuing one of those great gusty canine sighs. Then, slowly, he filled the darkened bedroom with his loudest House of Usher groan.

MUUUUUUUUUUUUUUAAAAAAAAAAAGH *(muff-whuff, fumff)* MOUGH.

Meaning, with embellishments, "I'm comfy." Tom returned and lay down. Briefly, we had peace. . . . A tactile hallucination in the form of red-hot sand drifted under the quilts, between my toes, into my throat, and onto various parts of my husband. We scratched and sighed. The dogs scratched, jingled their tags, and sighed. . . .

On a night like this, I know for sure that there's no such animal as the classic individual; that the enchantment of personality lies in each one's flaring-out, budding-off, from the primordial bond of our unity. Nothing brings it home like bad dreams when the hevrei are trying to sleep.

Back to translation. Itzik has a new address.

I. Bentov
10 Dana Street
Cambridge, Mass

5.27.59

Shalom Hevrei!

I don't have to apologize for not writing, because once you hear what troubles I've had you'll surely forgive me and more.

First off, I've separated from my wife may she live. Because we couldn't get along. I'm by myself here at a new address, I fixed myself up a nice apartment of one and a half rooms, actually the place is no bigger than half a room. But I'm contented, since it's quiet and no one hinders how I live. The child is with Miriam. She's a very sweet little girl already and I visit her every week (I see [Ora] tearing her hair). So you can imagine what it's been like on the home front over the past 6 months. . . .

In work, on the other hand, things have progressed rapidly, because it turns out you can't support two apartments and what-not on my salary. I had a fit over this and decided to look for work. I found a job but wasn't clear whether I'd gotten it. Meanwhile I went and gave notice. A panic arose. They said to me, What, silly, why are you leaving, stay with us as an outside consultant. In the meantime I was accepted at the new place, which . . . is ARTHUR D. LITTLE, the oldest and biggest company in the U.S. in commercial research. So I arranged for myself to be a consultant at both companies, that's good fun, but the best thing in the world is, I've become a professional inventor. . . .

I don't remember my parents' married life. Memories begin here, on Dana Street, with a story that my father liked to tell. My memory and his voice are alternately stirred within me, in plangent counterpoint.

I looked with foreboding upon a bread rusk, spread with cream cheese, like a flat stone between the prongs of his fingers. I told him that during Passover, bread was forbidden.

So then I am saying, "Little Pookli, don't worry, I have invented a special material that is God-proof. I have installed panels in the ceiling, he can't see through, you can eat your lunch."

My father and I never got to celebrate Passover together. All the tradition we'd had, when he died, was this cheeky tale of an inventor bringing unprecedented things into the world. Since his death, though, we've been suspended in Passover Eve; it's everlastingly two in the morning and the tablecloth is matzo crumbs, wine stains, spread-eagled books—it's past the middle of my life and I'm punch-drunk with philosophizing, hoarse with singing, and belly-sore with laughter, and too overtired to stop asking and answering everybody's tipsy questions: Who are we and who remembers? Who is God? Who put the bing in the walla-walla big bang? Who put the tribbles in the quadrotriti-cale? And who gets to clean up?

Shit, I knew I'd have to face this one again.

I. B. DEVELOPMENT CO.
407 Belmont St.
Belmont, Mass.

[date blurred]
Judging from the color of your last letter, it's ripe to be answered. I put your letters in a corner of my desk where they ripen until their color begins to fade and that's the signal to begin answering. . . .
Nu, we had a five-day honeymoon and got married, mazel tov, etc., and I don't know if I wrote you already, but I am now completely independent.

I mustered all my imagination and invented the name you see above on the stamp. A feat!! Life is a blast. I don't have to ride to work every day. I go down to the basement and I'm "at the office." I go down in the morning, come up for lunch, and work till nighttime. An average working day is 14 hours. I've bought all sorts of power tools, tools for woodworking, microscopes, chemical equipment, etc. I've even become a genuine bourgeois as I hire some Jew to come help me now and then. Sometimes [Sonia] also works down there on her sculpture, and then it's really good times. And besides Sharona I now have a bar-mitzvah-age son who it must be said is a good sweet kid. . . .

I hear that S—— went to [a kibbutz] and turned their heads so they're setting up a factory to make plastic toilets. Nu, nu . . .

Believe it. For Imre Tobias, aka Itzik Bentov, no girl child of a divorced wife could compete with *a bar-mitzvah-age son.* I could believe it, and despair. I think of Aia saying "You were his." I think of my missing pearl necklace.

And I try to think about Paul, Sonia's son from her previous marriage, who lived with my father until he went to college. It's not easy to think about Paul, having met him only twice. Once when I was six and barged into his room while he was putting on his socks. He cried "Hey!" and I ran. Once at my father's memorial service. Paul came up, introduced himself, spoke nicely. If asked to, I couldn't identify his face, or any feature of this member of my antifamily.

Hey, Paul! For all that you got to live with my father, you were an only child—did you too feel alone in the world? How are you, who are you? I don't have your address. I don't have the courage.

You with your forgotten face, whom I would have called brother.

. . .

And now: a resort hotel postcard.

The Empire Club
Hamilton, Bermuda

8.24.59
Shalom Hevrei!
So I flew the coop for a two-week vacation here. It's beautiful. A bit strange to see British police again and the British flag, but one gets used to it. . . .

It's a rum go, as they used to say. If the British police had caught my father with a Loretta under his coat, that spring of '48, he would have been shot. Yet his homemade weapon, and the struggle against the British, and the war against the Arabs, all connived to bring him on vacation to the Empire Club. . . .

I have reached the bottom of the stack: a business-sized airmail envelope. The Yaaris' kids, who collected stamps, razored off the upper right corner, but the return address provides a date. It is where my father lived during the seventies; his last home and laboratory. Inside the envelope, I find no letter. Instead, folded like paper, there is only a sheet, perhaps a sample, of flexible plastic.

Twelve

T ODAY WE MULCHED OVER the roses for the winter, and I noted it in the household notebook we call the Sweet Oak Diary. (Tom named our house for the chinquapins, or "sweet oaks," on our land.) Every day, I try to write something in the diary that will recapture the day when it's long gone. Before I came here, I traveled light; I lived out of five suitcases and couldn't be hauling memorabilia. But after our marriage, I said to myself: Enough! From now on, I will spree, I will gorge on memories. I will stuff our attic with clutter, our diaries with minutiae. I will dry flower petals and fill vases, and at the end of winter I will empty last year's potpourri into plastic bags, and label it, and stack it in rows in the attic, until my dying day. I will be a co-conspirator with my own mortality.

Maybe because the smell of change is in the air today, that airing of Nature's bed before she settles down to sleep, that fall smell that seems to vaporize off the iridescent yellow streams of corn pouring from the chutes of harvesters in the stubbled fields, fields almost ready for the magic glimmer of frost. A tantalizing, up-on-your-toes air, the air of change. I want to write some thoughts about memory, before I return to my story. I used to think of memory as the opposite of change. This was back when

I moved around constantly. Memory was the jewelry in my handbag, secure, immutable, outside myself: a memory was like a ring I put on my finger, or took off. It didn't change, and it didn't change me. Now that's over. Now it seems that in the stillness of my life, memory is change. It's like my Hemmed summer again: with each chapter I write, I knock on a new door. But the doors are all inside me. The Telling opens them. I don't know who I'll be at the end of this—if there is an end. . . .

Thirteen

STEADY RAIN. Outside, there's hardly a difference between the dark wet gray of the patio flagstones, in their frames of myrtle, and the sky through the treetops. I shut my eyes, because the person I'm thinking of isn't in the Passover photograph. Or anywhere on earth.

Commander No One lived in an old building on the corner of Shenkin and Rothschild, in Tel Aviv, and had agreed to see me very grudgingly. I arrived early, dodged two lanes of cars gunning their engines to reach the median park, and collapsed on a hot wooden bench, my handbag safely between my feet, in the shade of my denim skirt. The park was a strip of raked sand. Above me, the fretted roof of an arbor stretched away in shadowy colonnades that had stood for fifty years, and I threw back my head, willing myself to coolness in the lacy historic shade.

"Apparently, it is impossible to refuse you," Hemmed's former executive commander had opined over the phone. I had pictured a bored, lazy veteran, phone perched on his bulging undershirt. I knew his rank had been lieutenant colonel. When I said that my father had built Israel's first rocket, his boredom overflowed; he implied I was a fool. But with my referrals—Dr. Einstein, for starters—damn straight he couldn't refuse to see me.

By the time I rang his doorbell on the fourth floor, I was parched. The door opened a crack, admitting a glimpse of white silk, and an elderly voice said he'd thought I was due an hour later. Lt. Col. Gur, in a kimono, retreated behind the door, which he left open; his voice trailed back. Wait in the salon, help yourself to a book.

I stepped into another time. The plaster walls were amber with age; on a black walnut table, from deep in the Victorian era, stood a lamp shaded with tarnished lace. Two brocade-covered armchairs perched on bombé feet, beside a threadbare sofa with formal cushions, impossible to disturb. No one in Hemmed lived like this, I puzzled, sitting down. The blown hush of an electric fan caressed my ankles. The empty armchair faced me. Behind it, on the balcony, was abundant light, but little indoors. Faint steps, clinks, and rushes of water came from within the apartment. I smelled the woody must of old books, but saw only a small bookcase, in which the worn spines did not invite casual handling. Years of reading had infused the air to this degree, perhaps. Lacking a television, the room was hung with framed paintings at one and the same level around the walls. All were studies of water: pastoral streams, noble rivers, a fluid figure dancing in a blue shift; it was as though a canal surrounded the room, murmuring.

A good omen, I thought. I had special hopes for this interview, as Gur was the top of Hemmed's chain of command. I had, indeed, met Israel's fourth president, who with his late brother had been among Hemmed's leaders; but he preferred not to discuss details, instead giving me a valuable article that he'd written about Hemmed and Ben-Gurion. Details, however, were what I

was after. I unzipped the handbag at my feet and extracted a
notepad stuffed with a summer's worth of interviews. I loved
talking shop with Hemmed, and by now did it well. Was there
another American who knew why the English initials H.Z.M.N.
were sarcastically engraved on the first mortar built in a Jewish
workshop? But expertise was not an end in itself. I wanted to
revive my life; to write that dreamed-of poem about Hemmed
and my father, in which all the expert details would glow with the
sum of their meaning. It was my plan, and it was going nowhere.
I felt a lack within myself without letup. Maybe Gur would give
me the inspiration I had come so far to find.

Strangely, no one in Hemmed had mentioned how much more
Gur had done than command. He had spearheaded the National
Water Carrier, greatest of Israel's public works, which piped
water down the length of the country. He'd created the famous
"wall and watchtower" system for rapidly raising a kibbutz from
prefab parts, which had kept the Negev desert from becoming a
British base; my father's kibbutz, along with ten others, sprang up
there on the night of Yom Kippur, 1946. Gur had led the survey
of the country's northern border; he'd planned the building of
campuses, hospitals, museums. He was conducting discussions
with the Jordanian government about creating a chain of artificial lakes, each with twin names in Hebrew and Arabic, down the
eastern border. I learned all these facts later. During our memorable conversation, I did not know he was a one-man Israel.

Gur entered. He had cleaned up aggressively, wearing knife-
creased white slacks, a linen shirt, a tight shiny belt, and cuff
links. He carried a tray from whose rim the silverplate was worn,
and poured me a glass of water out of a sweating pitcher. I tried
not to gulp. My host lounged in the other chair, putting his feet

up on a velvet footstool; he had a lyrical build, thin, loose-jointed, with a quiff of white hair. Hardly military, even for the offbeat IDF; more like a retired dancer.

"Your salon is lovely."

He seemed bewildered, waving a long hand.

"A hundred years ago . . . ," he said. I prepared my notepad, a cue always met with helpful attention. Gur looked away.

"Are you a historian? I talked to a historian. What he wrote was boring."

"Actually, I'm a poet." He met my eyes; his were darkly lit, shaped like willow leaves. "I'd like to hear about Hemmed."

"About Genka," he sighed. I wondered why he began with the late Genka Ratner, whom I'd seen recently in a photograph that had given me a shock. Though at a second glance, Ratner was quite handsome.

"I came from a kibbutz in the thirties."

Gur paused, propping an elbow on the chair back; his fingers moved as no one in Hemmed moved, languidly. "We met in the Carmel gardens, where the youth of Haifa tended to meet. Genka was the most woman-chased boy in those gardens. He was magnetically attractive. His interests—calisthenics, photography—it was rare, in a backward place like Palestine. Compared to the rest of us, he was aristocratic. Then things changed. Some of us joined the Jewish Brigade, I went home to my kibbutz."

(*J.B. / W.W.II—'44?* I scribbled in my pad.)

"I heard that he worked for Shell Oil, went to England. . . ." Gur's teacup, on a saucer, rose to his lips; he replaced it with ineffable nuance on the table. "There was some whisper," he added, "about Genka's enlisting in the Jewish Brigade because he had been jilted by the second most beautiful girl in Haifa. Whom I knew. Also the first. But that, I believe, is untrue."

Good grief—love stories. Was this fellow senile? He must be in his eighties.

"I've read about Ratner's limpet mine," I cued.

"Yes. Then in 1947, I was invited to recruit young scientists. I asked Ben-Gurion, 'Who are my aides?' He replied, 'Genka.'"

"Ben-Gurion!" I blurted, then held my breath. This was history, the very beginning, the birth, of Hemmed.

"We met in my small apartment. We began working in my five-year-old daughter's room. On her little table, Genka began planning mines. Because my first task had been to take stock of all the weaponry we had in the country . . . and when I was done, *haya li hosheh ba'enayyim.*"

I glanced up. He brought his palms together and chopped sharply.

"We had nothing!"

I had darkness in my eyes, he'd said. In his peaked old face, I saw the darkness flowing out from his pupils. The past was alive there, and for an instant seemed to be still happening . . . to me.

"So the first thing was to make mines," he concluded.

"Was that how the Studio started, in Tel Aviv?"

"I made the Studio. The actor Binyamini had a shed on his roof. His ex-wife was a good friend, and through her, I got him to let us use it."

Echt '48! I gloated.

"Genka—was changed. By the war."

"I heard about his cursing."

"No, no!" Gur was stung. "You cannot!—it was not cursing! It was a type of poetry, Russian *poesia,*" he intoned, pointing didactically upward. To my delight, he launched into a story about four major figures: himself, Ben-Gurion, Ratner, and Slavin, head of Tas, a group that made weapons but didn't invent them. Early in the war, these four men met to discuss the Piat, a portable device allowing soldiers on foot, like the kibbutzniks in their fields, to repulse a tank. It was desperately needed. Ratner said that Hemmed could make a prototype Piat in a few weeks—

"If I have to use a hammer and stake," he vowed. Grim-visaged Slavin attacked him, pounding on the table. The Piat had over two hundred components, and Slavin did not think it could be built, without blueprints, in less than two years. Tas could do it in two years—Hemmed, never! Gur, who alone of these Russian-born men seems to have kept his temper, asked, "What if the Arab Legion doesn't wait two years? Their tanks are at Ramle as we speak."

"We will take our wives and children by the hand," morosely prophesied Slavin, "and walk into the sea."

The Hemmed higher-ups, upon hearing this, commenced the higher swearing of Russian *poesia*. Slavin fled the room. And the prime minister—when Gur remembered where he was, and looked—was smiling.

"Ben-Gurion appreciated *poesia*."

"I'll bet. And the Piat?"

"We had one Piat in the Studio. Someone smuggled it into the country. But no blueprints—"

"—right."

"We couldn't take it apart. How would we reassemble it? Genka said to us, 'Is there no roentgenography in Tel Aviv?' What today you call X-rays. He had a dentist, he made an appointment."

The commander and I exchanged security smiles. How like Hemmed, I was thinking, to do the unexpected thing. To bring a Projector Infantry Anti-Tank unit to the dentist, to be X-rayed. How very like my father.

"He was deformed, by the war." Inexplicably, the commander

began talking about Ratner's emotional troubles. Genka had shunned women, thrown black rages, and played obnoxious telephone pranks. Gur went into these gloomy anecdotes at length while I jotted, right foot jiggling, uneasily absorbed. Genka of the gardens, whose handsome face now wore the look of a cordoned-off site, almost eclipsed the story I had come for. Sometimes I tried to jog Gur back into the historical vein, without success, because life stories are as hard to redirect as watercourses. For Gur, nothing mattered more, after fifty years, than the way people in a crowded room had cleared space for Genka because he could not bear to exchange a word of small talk, because they forgave and loved him, because he was . . . himself. A certain mood began oppressing me, in which I felt that no matter how many stories I gathered, I would always fall short of truth. That my Hemmed hosts, spilling their memories, were no more coherent about the past than I was, and this effort to create worlds out of words was wasted time. That time did nothing but waste. . . .

What of the story I'd come for—the night bombings of Tel Aviv and Egypt's daily advancing tanks, the world of my father's rocket?

"People say the Studio was very exciting," I hinted.

Under his crisp shirt, Gur's chest rose and fell. He slid both feet to the floor and leaned forward, elbows on knees.

"I did not believe we would overcome the mass of the Arab armies."

I started to exclaim, but he shook his head.

"I denied this—for me, it was a question of facing . . . We had nothing. I would visit my child at ten at night; she'd hear me coming and stand in her bed." He was quietly intent, and again I had the sense that something was happening. To me. "There were no friends, no social life. You saw an attractive woman but you didn't register her, it was just a silhouette."

He didn't return my smile.

"So it was said of us that we were uncouth, we didn't appreciate women—you understand, you are a woman. We didn't even have time to apologize. We had no alcohol, no cigarettes, no food, nothing of—life, we sank everything, everything into work, all day and all night. In the Studio, to climb the stairs, I had to step over people sleeping on the stairs. They'd fall asleep on the desks, holding their pencils. That was the kind of pressure we had in that small group, who knew the truth and didn't tell it."

He held my eyes with one of his paradoxical expressions, alive as a struck match, yet far away, and altogether sensitive. I realized that Gur had given voice to emotions that Hemmed, as a rule—as the strict rule of their generation—did not. He spoke from their shadows. He was like a poet. Between our old-fashioned armchairs, as though carried on the fan's shirr, something communicated back and forth, between us. I felt it, and knew—this had happened. Two people had met; their souls were knit. I was stunned and confounded. For a mortifying moment I thought of what Gur's body would be, divested of the clothes into which he'd put his dignity, his masculinity. No, no, impossible, that comedown of wrecked flesh. That was not . . . I had too much respect. Yet how could such a knot be tied, and remain a thing of air?

I returned to the Hemmed story.

"But you won," I said softly. "You did win. Can you talk a little about your job as commander?"

Gur compressed his full lips, withdrew into his lounging pose; he sounded tired and elderly.

"I used to meet Sadeh at midnight in Café Cancan."

General Yitzhak Sadeh, I wrote, Israel's master warrior. In a café booth, a cap on his grand old cranium, with square glasses and twitching white beard. Gur opposite, young and wan. Sadeh sketched a planned operation, erased it, asked what weapons Hemmed could supply. Then Gur signed the paperwork. Yes, I could picture that. . . .

"Each morning I reported to Ben-Gurion, but we had no chemistry. The moment I walked into his office, he would do this—"

Gur grabbed the armrests and did something with his chin, so that fleetingly, an exact, malicious caricature of the prime minister glared at me. I giggled. Gur was disdainful.

"I told him the truth about the *gimmickim* the scientists thought up, which he wanted instantly, but I would never sign on anything we couldn't produce. Ben-Gurion believed in the mystique of science. I have no idols. . . ."

No idols.

"I didn't like the military. I am a civilian type, my friends were poets, actors, the *bohème*. . . ."

"Is Café Cancan still around?"

Gur winced. I explained the need for sights to help me to imagine the past. He looked away and his silence was a wound.

Oaf, I told myself. Talking about "the past," as if his life were something exhibited in a museum. He wanted nothing more to do with me; I was being evasive and on top of that, tactless. This visit was over. But I could not put my notes away, heft the handbag onto my shoulder, and walk out of his water-haunted room. The idea was unbearable. I could not go without receiving or offering a gift, unnameable since it was not sex. A mystery that would not let me leave kept me babbling, playing for time, until I asked a question unrelated to the story I'd come for. Ziva once said that life is a game of chance, but in Hebrew it also means "a game of destiny." A chance word flies from the lips, and destiny grinds its great wheel that never turns back again.

"You've told me a lot about Genka and the others, but not about yourself. Who were you, during that war?"

"I was not," he answered.

"'Not' . . . what?"

"I was not," he repeated.

Though the balcony was still ripe with sun, the room had dimmed. My notes were harder to see. The whites of Gur's eyes glinted as he looked up, and I sat very still. When he spoke again, it was in fragments, gleams. For the first time I was hearing what was hidden in his depths. His accent deepened; the word *v'* became a little song, *vehi-yeh*.

"I came from Russia in the twenties, when tens of thousands of Jews were trying to get out of Russia. They formed gangs— ordinary, middle-class people. My family was killed and scattered, I saw— You're alone, no family. I was seventeen. I left with a friend and two dogs. It was fall, we walked in the forest. . . ." He paused for a long time. "In short, we were contrabandists."

Contrabandistim. A word from a dead world. In its frozen twilight, a boy huddled under boughs of fir, wrapped in his coat-skirts. He had a pen, a watch, wolfhounds with slim maidenly muzzles, and one friend. The two of them trekked the Bessarabian border, crossed the Dniester, and made their way to Bucharest; they met up with Zionist youths, and Gur learned of jobs for cattle tenders in Palestine. With their dogs, they sat on a small freighter for three weeks, and landed in a strange new life.

"The first night we slept in a seaside hostel. The stars were incredible. In the morning we went to the Carmel gardens, and picked fruit, and ate."

Stars, I scribbled, *1st breakfast.* I thought it must have been like Eden, understanding why he grieved for Ratner, the handsome boy of the gardens. Meanwhile, light ebbed into the blurred tree-tops outside. The sinking sun drew my energy down with it; my hamstrings tightened, and a bubble of pain swelled around a headache. I had three buses to catch before home, supper, and sleep. Until my question was answered fully, I would wait. I was going on pure instinct, I knew these moments would mark my life. The noise of evening traffic reached us, a hubbub strafed by car horns.

"But why," I persisted, "why do you say that you were 'not,' in '48?"

"Do you know what a patriarchy is?"

I said yes. Gur disagreed. He did not think I understood. He'd been raised in a patriarchal family. He had known where he would be, how dressed, how occupied, at the stroke of every hour, until he occupied a reserved plot in the Leningrad cemetery. There was never the slightest mistake in the keeping of observances.

"I knew that Grandmother would kiss me, Grandfather not. I had a tutor in classical music, so my tastes would not become corrupted . . ." He broke off. "You're an American."

The balcony flooded with pink light that stopped at the doors. I rubbed my eyes. His voice was as distinct as it was soft, like printing on the nerves.

"When I walked out of Russia with my friend and our dogs, we came to a clearing, a bald hill in the forest. The sky was over-head, the forest around. We had to choose, on that hilltop, what direction to take, any direction except back. Suddenly, I was no one."

I was very tired, sustained by an unaccustomed feeling, *ven-eration*—but I still didn't get it. My commander had become a silhouette, fingertips set together in a cage; light traced each shadow-finger like a pale wire.

"When I fought the Russian gangs, I was one of them. I had to knife people, that's who I was. I was no one. When I joined the kibbutz they took my watch and my pen—it didn't matter, they gave me patched clothes and I was happy. People never asked where you were from, or what family, so I didn't know who I was. I painted, wrote poetry, sang—why not? I didn't need permission from God or universities. Suddenly you are no one. You can do anything."

I scrawled sentences I couldn't properly see to form, for some time after he had finished speaking.

Finally I had the story, though not the one I'd come for.

Hemmed had created weapons from scratch, "from nothing," they always said. Now I knew where their work began. Not where I'd looked: not in the Studio, the Rise, the Weizmann Institute of Science, or Rafael. In the old countries of origin, their families, their homes, the roots of belonging were erased. This was their generation's story, the unscathed as well as the orphaned, because violence against a group strikes every member: it acts upon the common mind. Behind them, their roots were gone. Before them, armies stood. From the ground zero of identity, they'd hurled themselves forward in every direction except back. Abandoning the past, they were no one, they could do anything! Israel rose out of their generation's momentum as a sand devil twists up from the sand into a spinning tower, mimicking the spinning clay on a potter's wheel, though inside there are no shaping hands, there is nothing. Before they'd invented a single weapon from scratch, Hemmed's work had begun with inventing new selves, from the nothing within themselves. That was the source of their explosive energy—and my father had been one of them. He'd shunned the past, living for his next horizon, his next idea.

He had been an inventor.

He had been No One.

I wrote rapidly in Gur's salon at dusk. Right below my breastbone a pressure was expanding, words crowding so I could hardly keep up with them. Now, long after the commander has died, I can identify that pressure. The child who left her father's car one summer evening, knowing what she was never to have; who vanished from memory between one minute and the next. She was the lack within myself, the voiceless void. I had gone forward into the person I'd become, itinerant and restless. I was No One. The commander gave me my proper name. As I sit writing here, bestowing this name upon a divided self—a child too disillusioned for memory to hold on to, and a woman pursuing a dream—I feel the same welling-up pressure full of words that I did long ago, in Gur's sad and lovely room. It is the convergence of past and present, the divided self joining in pain and joy, in waves of darkness and light.

"So to answer your question," the commander said from his dark armchair, "I have no special memories; just impersonal memories, like photographs. The weight of my time and energy was given to Hemmed, but not the weight of my life."

I put down my pen and flexed my cramped fingers, wondering how to thank the old man whose willow-leaf eyes, I knew, were meeting mine. There ought to be a protocol expressing thanks for a gift of spiritual magnitude. There had to be, but I was an American.

"I think," I said, "that you've told me very important things. I really am grateful."

"I've opened myself to you as I have not to any woman," Gur said, a bit uncertainly, "in many years. You will not write these personal things, please? Just the Hemmed story."

"Of course, as you wish." I got up stiffly, with a sigh; closed the notepad, and hoisted my handbag onto my shoulder. Gur escorted me to the door, courtly as ever, beginning to wander a little in his remarks; enough to reveal that he'd given, as he would say, weight, to the effort of our conversation. I thanked him again, clasped his long, lumpy, cold hand in both of mine, and said good-bye. As I pressed the button to light the stairwell, I saw the edge of his door closing. I felt suddenly numb and light.

The evening on Rothschild Boulevard was brighter than expected, well lit by streetlights and cheerful traffic, lights everywhere; blue-white windows, red and yellow taillights, orange sodium lamps. On the bus stop shelter, little bulbs framed a poster in which a bikini-clad model appeared to solicit a giant snack food. Around me, people waited for the express to the central station: a woman my age, comfortably loafing in a loose black shirt and slacks; a young boy with a silver earring and a German shepherd, a girl in a thin cotton dress, pacing slowly, her brown tresses blown about her waist. The sky was a hearty purple. My ride home would not be so bad. Then the sound of my own name caught my attention, coming from above. I turned in a circle, looked up, and saw the white quiff of Gur's hair on his fourth-floor balcony. Foreshortened to elbows, he hung far over the rail and cupped his hands to form a trumpet. I had to gesture encouragingly, several times, until at last I heard what he was calling: "If you write any poetry, send me some!"

Part Four: What the Mind Does

Fourteen

"DADDY, ARE YOU SURE?" I kept asking him through the late seventies; it never occurred to me to think that he was going mad. "Are you sure, Daddy?" I deliberately un-knew, and un-knew that I knew. He was working on the Transcendometer. It was not a weapon, it was worse.

Go into any synagogue when it is empty and dark, and silent. Approach the *bima,* the dais at the front, the original Dais Theater. Behind the lectern is the cabinet of the sacred scrolls, shut now, and richly curtained. But look up. Above—guiding your steps—hangs a lit lamp. A few watts shining. It is a reminder of King Solomon's temple, where one oil lamp in the golden seven-branched candelabrum was kept burning, in order to light the others. But you, Jew or Gentile, already knew its meaning. You whose ancestors domesticated fire, you know the Perpetual Flame.

Can you imagine what might turn this light, in a human heart, into a destroying force?

My father traveled all over in those days. He flew to Haiti, to witness bloodless surgery performed by a sorcerer whose only instruments were the passes he made with his hands, and colorful rays from an occult source. He flew to Japan, where some researcher took an intense interest in him that changed, overnight, he knew not why, into an intense objection to him. He flew to Esalen and Banff, where brilliant people congregated to transform the world through psychic power. Names filled his talk, and every name was haloed with eerie accomplishment—he was a friend of mediums, prescients, healers, the new, more highly evolved human beings; in short, nobody I was equipped to judge. He flew to Switzerland, where Guru Mahananda was reportedly teaching people to walk through walls and fly. He flew (by plane) to Israel, where the celebrated conjuror Phelps was being investigated at the Weizmann Institute, to see how he untied shoelaces with his thoughts. From this last expedition, my father came home raving.

For months, I have been putting clues together. Last September, I did not guess that the Telling would bind me to the tale of my father's madness until I had an explanation for it. If you had asked me, in September, I would never have dared follow him that far, to know his mind in its collapse. Yet the clues have accumulated, and maybe I have changed enough—been changed, by the Telling—that I can see a pattern in them without fearing for my own sanity. Ever since I discovered that I was No One, the forgetful, I have been turning into someone new. Someone who resets the joints of time. Someone, the Rememberer. And this is what I think happened to my father.

On a grassy hill stood a mansion, a museum devoted to the founder of Christian Science. Behind the museum, which was open to the public, a greensward sloped down toward a reflecting pool, square, green, ornamented with a stone angel. The angel's feathers were shaped like butter knives. A poor sculpture, my father used to say, but a beautiful spot. We went there on our Saturdays. We would sit on the stone steps that led beneath the surface of the pool, turning to steps of dappled tourmaline. My father talked for hours, in the smell of mown grass and the scent of water. An aspen leaf twirled slowly on the ripples, and sunlight tangled in radiant knots.

"Daddy," I asked, my head spinning, "are you sure?"

"Pookli, they are dangerous, evil, low-down types, real bastards." He was chattering at breakneck speed, had been for hours. Telling me how he'd been attacked, in his own home, while he was meditating and traveling out-of-body, by the alien beings who possessed Dennis Phelps.

"They attacked me and I could not get back into my body, I was scared they would take me over like they did to Dennis. They are trying to take over all the sensitives. They are very powerful. Never have I had such an experience before at the astral level—there are all sorts of riffraff hanging around there but I always tell them Scram! and they run away. These were like jet fighters. I had to escape to the mental plane where they cannot follow, and then I am creeping back to my body, slowly, slowly, and bang! Twice they caught me and I had to run. It was terrible. Meanwhile, while my body is trapped in the chair meditating, the bastards are flying around the house, throwing things. Sonia had to dodge the things they were throwing and she started to scream, and they made a horrible rotten smell, it takes forever to dissipate. I can still smell it."

His neck was too thin, the skin over his Adam's apple was ridged like a parsnip. He'd had all his teeth pulled, as if to avert some dreadful prophecy, and his new artificial teeth were square-edged, chalky, when he licked his dry lips.

"You're saying Sonia saw things being thrown at her? Are you sure she wasn't just—upset?"

"Plates, plates were flying through the air and hitting her! Naturally she was upset. The stink! I cannot—this has never happened before, in my experience. It is terrible, Pookli, I don't know how to fight it."

I knew about alien possession. In my experience, it happened on *Star Trek.*

"Did you ask Mahananda about it?"

"Mahananda is surrounded by stupid moron business types who will not take messages."

"Well—is there anyone who can help?" I wanted to help. I couldn't travel out-of-body to defend him because I wasn't evolved enough. I wanted to cry. My father turned, gave me a look; his green eyes were bloodshot, the pupils mere pinpricks.

"*Nu,* Pookli," he said with his old humor, "I have extrrricated myself from vorse situations."

"Daddy, are you sure it wasn't a dream?"

"I am telling you," he said, omitting my nickname. "I am just escaping with my life."

He had never fit in. As a boy he'd been outlawed and had fled the fascist state. As a young man on the kibbutz, he'd been the odd-ball, the *astronaut.* Everyone laughed at his jokes but he could not fit, no matter how fast he hauled cement hods or cleverly rejiggered the tractor. In Hemmed, he'd been the bright mechanic without a college degree, uncredentialed for Israel's new scientific establishment. During his brief stay at Rafael, the national defense research institute that Hemmed became, he'd wanted to get out of making weapons. In the United States—ah! There he came close; he was Invention-a-Minute Ben, a successful entrepreneur. But this man who sang constantly as he worked, as he drove his car, as he shopped—booming "Camel, My Camel" in the automotive aisle at Sears—never sang an American tune. I hate to say it: my father never loved my country. It was a marriage of amazing convenience, for them both. His America was a home laboratory; his countrymen were a small circle of techies, his wife was Russian.

I believe this made him wistful. He used to joke, "I am just an old-fashioned Yahnkee tinkerer."

It didn't help. He still didn't fit.

Then he published *Stalking the Wild Pendulum: On the Mechanics of Consciousness.*

Suddenly, as if by magic, he was transformed from a hi-tech hermit into a semi-celebrity surrounded by the zany warmth of crowds. People really dug him—him, not just his patents. Whole choruses of people told him that his ideas were their guiding light. That the touch of his hands cured their diseases. That his laugh restored their joy in life. Women, for once, wanted to seduce his mind. He, Itzik Bentov, was their hero. From all quarters, bouquets of friendship, admiration, were thrust into his arms, under his feet. Indian gurus swung their wise heads in figure eights when he spoke to them. He fit here, in the New Age, among teleports and telepaths, psychics and channelers, in the merry old land of Oz.

Itzik Bentov was Someone!

He began to remember. He was becoming Someone, the rememberer. . . . One sunny day, on the Boston Common, as we were feeding the ducklings, he turned to me with the smile of a very young man.

"You know?! I used to have a little sister," he said. His sister had died of starvation in a death camp. I was twenty-one. He had never mentioned her to me before.

He was opening up to his feelings, and his memories . . . and out of their secret places surged forth monstrous terror, hopeless sorrow. The Holocaust with all its abominable horsemen and chariots charged straight at him, alone on the field of memory, without weapons, without cover. All that he'd never allowed himself to feel, reflect upon, or remember went rampaging through him, uncontrollable.

He dreamed, perhaps hallucinated, that he was being attacked by aliens. Figuratively speaking, this was true. A famous author once wrote that Auschwitz was another planet.

He ached with sleeplessness, complained bitterly of it. The round of international conferences and lectures fatigued and jangled him, without paying enough; then at home he could not sleep because of his anxious work on the Transcendometer. Even before his extraterrestrial SS slugged him, meditation had failed to bring him ease because he flew out of his body—as exhausting, he swore, as jet travel. Nothing could slow him down, nothing could calm him. He segued into paranoia. Once again in his life he was RESTLESS . . . as if he were doomed to be the wild pendulum, flickering on and off, on and off, continuously blurring across the universe.

He took a powerful sedative, Stelazine. He told me that it had been prescribed by a New Age therapist, whom I remember from the day of the memorial service. Afterwards, when I knocked on my father's front door, the therapist opened it; he looked silently in my face. Years later, reading a newspaper article, I learned that this doctor had surrendered his license to the state Board of Registration in Medicine, following reports that he had sexually abused his female patients, sometimes after sedating them. I don't know what became of him or of the reports. I don't know what my father, who in his medical work sought to save lives, would have made of it. But evil preys on the distracted soul, and my father's Transcendometer was a consummate work of spiritual distraction. Though his death was accidental, evil presided over his end. So I felt, I recall, upon seeing that doctor's signature on the bottom line of my father's death certificate.

Sunlight tangled on the reflecting pool. A few needles of white pine floated close to the edge, above their drowned and swollen shadows. My father was intrigued.

"Look, the pine needles, they are casting round shadows." He peered into the water, trying to figure out why . . . so our afternoon passed, until it was time for the leave-taking ritual he had invented that same year. This ritual is my most precious, yet saddest, memory, because of what it reveals. My father's madness was triggered by the very thing that could have healed him, had he lived long enough: the opening up of his feelings.

He would cup my face in his hands and say, "A real daughter! I have a real, grown daughter! One hundred percent genuine materials, nothing synthetic!"

I was his. We were a family. He said so, in his way, at the last.

I'm going to look the devil in the face.

I step out, and return to my study lugging a stack of brown notebooks with MIT labels. On each label is my father's handwritten title: *IDEA BOOK*. His MIT friends probably gave him these for free; despite having no college degree, he regularly consulted with engineers, scientists, and medical doctors. I hoist the notebooks with effort, spring bindings cutting into my palms, and drop them on top of my desk. The stack looks tall. But also short.

Here is the entire career of Invention-a-Minute Ben, the father I knew, or thought I knew: his laboratory notes for everything between 1960 and the year he died, from "Instant Ice" and "Inflatable Upholstered Furniture" to the "Appliance to Support Gums Around Teeth," the "Slow Motion Motor," the "Cardiac Catheter"—his masterpiece—and finally that devil's work, the "Transcendometer."

I stroke a brown cover, its glossiness dulled around the corners, where the cardboard is like felt. The page is yellowed, blue lines turning green. When I sniff, the chemical fragrance of his ballpoint ink is alive on the paper.

I turn to the lab notes for the Transcendometer, and immediately see something wrong. On this page, the writing has changed. It doesn't flow in orderly rows around beautiful pen drawings. It's a chaos. A wild scrawl in which phrases stick out, circled like thoughts in a comic strip: "Vitality" is one, "Reynolds Process Cambridge Plating," another. These bubbles give the pages an effervescence, a lack of focus. Exclamation points abound. Long lists stream slantwise across the page, and they include about every common material my father used: brass tubing, copper tubing, fiberglass, resin, Lucite, Masonite, foam rubber, Saran, glass, plastic.

Mainly, the Transcendometer was a chair. I believe it was an ordinary folding chair such as you'd find at a card table. The picture that emerges from my father's notes is of himself, seated in this chair. Bolt upright, eyes shut, engaged in a meditative discipline whose name you may have heard, but which I'll avoid because it's copyrighted. Two feet away hovers a sensor, on the end of a long rod like a microphone boom. Should my father grow tired and his head list backward, it will be braced by a padded headrest unpleasantly recalling the manner of his death.

This simple setup produced a furor of experimental tryouts, listed below:

- With the blue pillow
- Without the blue pillow
- In a wet lab coat
- Rigid or limp
- Breathing or not
- With head shaking back & forth
- With sponge rubber on seat
- With polyurethane foam in a slipcover
- Layers of sponge sandwiched between layers of poly
- With a board on the seat

While the chair and the man in it were being wetted, shaken, foamed, boarded, et cetera, the sensor was moved to within two feet of various areas of my father: back of head, heart, top of head, forehead, solar plexus, navel. In each location, it picked up signals. A graph shows two types of signals, electromagnetic and ballistic, charted against the physical areas producing them. The body parts are labeled with symbols: the knee is a square, the heart is an X.

· · ·

There was nothing untoward about my father's experimenting on himself. He'd made devices that had left orange scabs on his scalp, and boasted that for the purpose of testing electrodes, it paid to be bald. He preferred to try devices on his own body when it was practicable to do so; he even confounded a dentist by taking over in the middle of his own root canal. "Gi' ee gah mir-r'er!" he'd commanded, grabbing the dental tools. What unsettles me about these experiments is something new.

For starters. From its inception in 1973, the Transcendometer was inhabited by a perverse spirit. It was sensitive to what was not there and oblivious to what was. It produced "STRANGE PEAKS." It had no difficulty sensing a person twenty feet away, but only the first time; thereafter it didn't care. Waving hands at it also caused a signal the first time, but failed to get a rise out of it afterwards. The Transcendometer's whims mystified my father, who scrambled to win its consistent attention. He wrung his brains, he went hog-wild. His records are a maze of short-lived triumphs:

- THE SIGNAL BECOMES LARGER! SPIKIER!
  THE BEST YET!
- WE ARE USING NOW MY LAFAYETTE
  OSCILLATOR!
- THIS IS THE BEST ARRANGEMENT!

Then, suddenly,

- WHY?

All machines, never doubt it, manifest personalities to their creators; my father's mechanical progeny usually befriended him. I'm not saying that the Transcendometer was one of your rogue robots; still, there was something about it. My father's plans for his invention were benign, and I'm not saying that he was Dr. Frankenstein—but he had created a bitch.

What exactly was it supposed to do? I turn to my father's cogent, two-page proposal. *We have at present*, begins the typed document, *a device which can measure and outline the human aura.* He seems to have been measuring electrical fields; I'm not qualified to assess his findings. Besides revealing the chair's true purpose, he proposed three clinical applications.

1. The device could measure the force of the heart's con-
traction.
2. The device could tranfer its scrutiny from the heart to the mind. It could measure the difference in frequency between random thoughts and focused thoughts. He had in mind a kind of creative thinker's rhythm method—to predict when ideas might be conceived.
3. The chair might be developed into an instrument for detect-ing anxiety, something akin to a lie detector.

Not one of these clinical applications held my father's gen-uine interest. This was unprecedented. He merely wished to sell a plausible tool in the medical market and use the funds for his real research. Till now, when he invented a thing, he'd always had a goal that was clear and maybe a little impossible. Never before had he set out with three commercial window dressings, and one major goal that was impossible not because the technology did not exist, but because reality, as we define it, did.

The Transcendometer's true purpose thrust it forever from the company of all the other machines he'd made. It was not created to solve an earthly problem; not to help, fix, heal, or repair. Not even to destroy. It was to carry out an eldritch investigation. In his own words, *I see this tool as a major development which may shed light on the otherwise fuzzy and contradicting claims of the people involved in the study of the paranormal.*

In the Transcendometer's fifth inconclusive and unprofitable year, my father made experimental subjects of his wife and friends. Their signals are preserved on gridded Parke-Davis recording tape, folded into a wad, and secured with an elastic band. Hundreds of yards are tucked in here. It's odd to think of those twelve or thirteen folks in my father's . . . sect . . . sitting in his chair, eyes shut in meditative trance. A signal processing unit and a strip chart recorder ran, while each subject boosted his or her mind into the highest of the invisible realities, Cosmic Consciousness—or God. That was how the Transcendometer got its name. It was meant to detect, and measure, the transcendent moment of the mind's union with God.

My father pictured God along Eastern religious lines, as pure consciousness, in absolute rest, beneath the wavy agitation of the material cosmos. God was Love: love without emotions, love without bodies, love without history, love without war. The blue becalmed sea beneath the universe. He pined for that. He ached for rest. He wished to soar. Who will give me the wings of a dove? To soar to his rest in God. But he was an inventor, a technologist; so following his natural bent, he invented a device to boost him toward God. Icarus, however, is not his story. He did not fly too near, or anywhere near God. He flew into the trick horizon of an illusion, believing that he could reach God through mechanical understanding alone, through the knowledge and use of tools.

It was the same error he'd made believing that evolution worked just like a conveyer belt. It stemmed directly from his denial of emotion, and his rejection of the body. It followed from his ranking all the different types of consciousness in the world from lower to higher, as you grade sandpaper, in hopes of defining and reaching the top grade, which was God. Yet I would swear that his arrogance was not Icarian, it was unintentional, he didn't know any better. He overreached out of love. His love was immature, greedy as a nestling's, but my father loved God. He might yet have found another way. He might have reached God through facing and naming the void within him, through finding out that he was No One. What monsters denned in his depths, though, with what corridors of teeth! No wonder that if he sensed the void within himself, he desired not to know it, but to fly, fly, fly. . . . Itzhak Bentov loved God. I would swear to it. I would swear that he was destroying himself with the purest flame that burns.

I am unwinding the recording tape, inspecting the output of his Transcendometer. What can I make of the human mind's merging with the mind of God? Most of the ink tracks that record and measure it resemble the tongue-and-groove sutures in the bones of the skull. A few, the zigzags of a sewing machine mishap. . . .

Had he only lived, to survive the ravages of memory and know himself better, to be reassured by loved ones—had my dear father only lived. Random chance killed him. Chance, which saves a man from the death camps, which saves him—for later.

The killer explosion, the spark in the void, the condition of my existence, the risk at the root of the world. Random chance. It's Passover Eve, so let me pose a question: Against what should I rebel?

I stare out the bay window of my study. I see the winter perspective of bare oaks and ashes, their trunks becoming finer and denser with distance. Heavy mist makes the woods stop short, as if before a clouded mirror. The mist is dark silver. And blessed be the name of the Lord. Fat drops hang in the mist, streaking the glass panes where they touch.

Fifteen

I F COMMANDER No One gave me a name, it was this man who gave me back my faith. Like his commander, he's not to be seen in the Passover photograph. Yet as I regard the group from the vantage point of their photographer, who would have been stepping backward, calling out orders and jokes, then holding a breath so as not to jiggle the apparatus, I keenly sense Matti's presence. That personality of his. It's in the manner of their drawing together, energies poised to erupt. Time itself seems about to burst into a laugh, filling the distance from that antique shutter's click until the present instant, the next present instant, and the next—keep counting!

At the end of our conversation, when the door shut and I stood on the landing of a top floor, on the summit of a mountain, I hated to go down. From a shutter above the landing, sun streamed in stripes, gliding across the wall of the descending stairwell. I remembered how, hours before, I had lingered here, listening to a violin playing an arrangement of a Bach cello suite. When the doorbell buzzed, it stopped. Dr. Amitai Geller, alias Einstein, appeared, his bow descending to his side.

The retired head of the Luz missile program.

Now I saw the curved shine where a mop had swept each

beige stone step. There lay my road, down into the flatlands where I had lost my faith. Would I be able to keep what Matti had given me? I belonged up here, not down there among the half truths and petty games, the deadening constant effort to imitate the norm, oh God the dungeons of normality where I'd been trying, ever since my father's death, to fit. I had forgotten what it was like to feel completely at home until I'd talked with Matti. I stood thinking. I would camp out here. Sit on the stairs, wait for something.

There was nothing odd about his salon, unless perfection is odd. In this room filled with mountain light, where everything seemed both new and everlasting—folds of lace falling from the round table, a fluted glass vase with white petunias—I kept discovering things that I'd thought I'd known, but understood for the first time. Matti sat by a window, composed, in a polo shirt with a collar that matched his eyes, and twill khakis. He hugged himself as he talked. He seemed doll-like, with his flaxen hair; only an intangible sort of ether about him meant the very opposite of a doll, the far-yonder of the human scale: a genius. He could say very surprising things.

"The Germans had such an imagination for rocketry!" His blue eyes, I noticed, had strange reddish flecks. "If only, if only Hitler had not forced the Jewish scientists to flee, German rocketry would be the most beautiful in the world!"

He had joined Hemmed for a common reason, the same that had led him into engineering: "I wanted to do something for the country." Like everyone else he had paid something for his idealism. Modest about his key role at Rafael, the national institute of defense research, he kept up a mildly boastful refrain about the position in the Israel Philharmonic Orchestra that he'd been offered and turned down, for Matti would always regret the musician he might have been. For him too, the music-playing arm had been detached from the gramophone. . . . No, it was not Dr. Strangelove whose door was shut behind me, at the top of the stairs that I could not, I could not descend. It would be like stuffing my feet back into a pair of crippling shoes. But I had no right to stay, or even sit. I grieved on my feet.

I sat at his table, with my notepad, asking Matti about his experiences in Hemmed.

"In the Studio, we didn't know anything. Everything was new to us."

"Such as?"

"The Monroe effect was discovered by burglars in London."

Huh?

He elaborated in a tone of hushed excitement, didactic, antic.

"They burgled safes by blasting them with explosives inside a hollow brick. The police would find the brick manufacturer's name stamped into the safe's metal door, and this was the negative, you see, of the name that was raised in relief, on the brick. *So.* Explosives make dents in metal. Explosives with hollows in them make deeper dents. Hollows focus the shockwave of an explosion the way a lens focuses waves of light. For military purposes, the hollow could be lined with—"

"Brass!"

"Soft copper," said Matti, not missing a beat. "Which shoots out at five to six miles per second, like a molten bullet. Thus copper, which is soft, can actually penetrate steel armor."

"Oh. . . . Now I get it!"

I had not been idle about researching the Hollow Charge.

Using Aia's referrals, I had visited a housewife suffering a nervous breakdown. She'd remembered my father shooting Hollow Charges at a derailed British locomotive, but didn't know with what kind of gun; after raising ten children, she had cried, she couldn't even remember her own telephone number! I had talked to a surgeon, profoundly depressed that he wasn't an engineer. He had once shot a Hollow Charge, a dud, and remembered absolutely nothing about it except the noise it had made, which I duly recorded: *Ploof.* Thus far, my research had given me a better picture of Hollow Lives than of Hemmed's most beloved device.

"Matti, this is the first time I've really understood the Hollow Charge. It's all about waves, isn't it? Focusing the shockwaves?"

"And the liner! At first we ignored the copper lining, that's why Racah's Hollow Charge failed."

"The big one they tried to blast into the Old City with?"

"It didn't even scratch the wall. Later we began making small Hollow Charges with liners, which did better. Your father worked on the Loretta because we needed a gun that could shoot a large-sized Hollow Charge. The Loretta did not shoot very hard, *ja*? Thus we required a bigger, more powerful charge, to make up for what the Loretta lacked."

"Matti—" I sat bolt upright. Suddenly everything fit, it was wildly obvious. And obviously wild. "Matti. Are you saying that—the Loretta was built to shoot the Hollow Charge—and—*that was how Hemmed planned to stop tanks?*"

His blue eyes reflected my surprise.

"The advantage of the Loretta is no recoil, *ja*? Antitank guns are usually mounted on heavy vehicles that absorb the recoil without flipping over. We had no such vehicles. The Loretta could be mounted on the jeeps we did have."

"Let me get this straight. Hemmed created two experimental weapons, the Hollow Charge and the Loretta, from scratch, then combined them into one big experiment? So this whole ballistics program was a kind of gamble?"

"*So.*"

LORETTA + HC = ANTITANK, I wrote, and let my pen fall. I started to laugh inside, the soft laughter of discovery.

"Matti, why on earth was it called the Loretta?"

"I called it that! After the beautiful Loretta Young!"

Listening to Matti was an adventure, I never knew where we'd end up, only that I'd be laughing. He showed me his article in the *Journal of Irreproducible Results.*

"It is absurd, it is a joke," he warned Teutonically. "This is what the army, any army, wants from a weapon." I copied the joke:

A. *Cost = 0*

B. *R & D Time = 0*

C. *The Weapon Does Everything*

D. *The Weapon Is Invisible*

"I always tell the young engineers, the reason for these specs is that it costs the army nothing to make demands," said Matti snidely.

"Did you make this invisible weapon?"

"*Ja.*" He grinned. "More or less. Do you know how the Inuits of the Arctic catch a whale without tools?" He told me. It was elegant. "But the compromise with a perfect weapon is a weapon that is faster, cheaper, and does 90 percent, leaving out the 10 percent that makes the rest too expensive or too slow. America has a problem with overdesign, because you want global capability, you want weapons that work in the jungle as well as the desert. In Israel, we design only for our immediate environment." When Matti discussed his field, he mentioned neither Bechtel nor Intel, but the toy manufacturer Mattel, from which he'd learned to use mock-ups of the final product.

"Most engineers don't believe their designs will work," he explained, "so you build a mock-up, a toy, just to see it. You have to believe the thing is there. You have to get an overview—to get it, you need an open mind, open eyes, to have experience—and it gives you your internal world." We pored together over a book about Rafael, with photographs of Ben-Gurion, Shimon Peres, and Moshe Dayan at weapons tests along with Matti, Genka, and others identified only by nicknames. In the fifties, Israel, already poor, had doubled its population by absorbing refugee and immigrant Jews from Arab countries; with its resources stretched thin, it could not afford separate programs for sea-to-sea, air-to-sea, and ground-to-ground missiles—all necessary for a country with a long coastline and multiple hostile neighbors. The Luz was a multitasker: in various versions, it did all three functions. "We were the first to develop solid propellant sustainer motors," Matti proudly confided. Every rocket, he said, had a sustainer motor with propellant, which sustained the flight by producing gases, and a booster for short-term effects. What impressed me most about the Luz's story was the staggering number of to-do lists. I won't even try to describe the organizational push needed to begin building missiles in large numbers, instead of prototypes in Rafael's old machine shop. Merely for a simple test, say, at sea, there were hosts of puzzles, all of them new: (1) finding the moment of best visibility to launch the Luz, (2) its storage and (3) maintenance and (4) stability on board ship, (5) whether Lieutenant K. would be drowsy from the pills he took for seasickness, et cetera. And those puzzles, concerned with launch details, were what Rafael had the luxury of solving after managing to coordinate—for a single test—(1) the project's technical staff, (2) the photography unit, (3) the communications equipment and personnel, (4) the medical unit, (5) air force crews and (6) supporting aircraft, (7) the ground crew to prepare ad hoc runways, et cetera. . . . At last I understood the unexpected reply of one of my interviewees, a Rafael engineer, when I asked him what Hemmed's greatest contribution had been: "Project management!"

Matti and I examined a genealogical tree chart of Israeli missiles; the Luz and later projects were depicted as branches, while the base of the tree was Hemmed's 1948 missile, the Shark. It was made of wood. If I understand correctly, a jeep engine and a snorkel were involved in it somehow.

"The fact that I am alive surprises me," Matti declared.

"That sounds like a nice frame of mind."

"Maybe! I mean I nearly died many times during that period. Once I was out in a motorboat with a friend, and I dismantled a British flare bomb we found floating in the sea."

"To take to the Palmah?"

"No, I was just curious. Most of the dangerous things I did were for curiosity."

"About undetonated bombs?"

"I put the glass flare bottle in a bilge bucket. Then my friend tossed his cigarette away—"

"No!"

"It rained glass, right in my eyes. I went blind. The doctor picked out the glass splinters with a tweezer, one by one. He didn't think I'd recover. I was only seventeen. After a month I saw light, and gradually. . . . All because of curiosity."

He was looking at me through resurrected eyes; those red spots in them were scars. The blue flesh of vision had been pierced, torn, stretched. Matti's look grew quiet and fond.

"You have your father's laugh," he said.

I could have told him that I hadn't laughed this way since my father died.

How many steps from his shut door, to cross from the present moment into the memory of it? I could still feel the vibration of Matti's voice in my shoulders, my hair. Down there, in the flat-lands, no more laughter. Maybe I really would sit a moment. What harm could it possibly do? I needed to think about my rediscovered faith, what it meant now, how I owed it to the Coanda effect.

If water runs down a pipe and meets a spike, what happens in the moment before it chooses to flow onward? Nobody knew. After the war, Matti was sent abroad to study. At Stanford he was assigned a problem: not to solve it, but to analyze the various limping solutions that hadn't licked the Coanda effect. If water runs down a pipe and meets a spike sticking up the middle, pressure on one side or the other will make it flow left or right. But the choice does not happen instantly. It takes time. What went on during this certain period of time? No one knew.

Matti labored on his problem until he knew it as intimately as the violin scores he played in the East Bay Quartet. The Coanda effect was a monster. He decided to flee it and traveled with his wife, seeing sights, taking pictures. He came home to the Coanda effect, which remained unaltered by the Grand Canyon and Yosemite, as if nothing ever happened in the world except water creeping down pipes. One night, instead of working, Matti collected all the boxes of slides from his travels and sorted them. Tired and replete with images of elephant seals and ghost towns, he fell asleep on the couch, and dreamed. This was Matti's dream. He was a stream of water flowing into the mouth of a pipe. For a prolonged moment, he pooled around himself, looking back over his liquid shoulder. He observed a silvery bulge, a bubble, swelling between his skin and the iron pipe wall. He watched the event unfold, and woke, eighteen minutes later (he noted the precise time), around a great bend of his life.

The next day, he showed a few pages of mathematical description to his adviser, who said, "Very well, we now have a paper to represent Stanford at the annual fluidics conference in ten days."

"You are joking." Matti was horrified. "I have not written a paper. I have just written some hand calculations, and not very precise, *ja?*"

"Ten days," replied his adviser.

"The most amazing part was this!" Matti told me. "I turned the dream into a story." He punched his mathematical Telling—the story of water as if it had been himself—with two fingers, into an old IBM typewriter.

I wrote down the story in a script so suddenly enlarged it made the rest of the page look puny. Matti was sane. And his math worked. His vision was not an illusion. The vision was not an illusion!

"My father had experiences like that; he said he was visiting other realities. What do you think?"

Only I didn't ask him out loud, I was afraid to. Then Matti told me how it was with people who are afraid. He'd returned home wanting to build guided missiles, at a time when Israeli experience was mainly limited to the wooden Shark.

"People were afraid to try metal. I was the first—among the first, to say that missiles could be made with casting and so on, from a single piece of metal. But I had to educate the system. People thought missiles had to be sheet metal, like planes. They were afraid. They had to check everything. That's how it is with people who are afraid. I tell the young engineers, 'If you believe you can do it, maybe you can. If you don't, you surely can't.' You must see the goal before your eyes. Then make lots of mistakes."

"Mistakes?"

"Mistakes are not the end of the world!" Matti's soft voice was deeper than you'd expect. Now he laid his tight-fingered hands on the table and spoke with passion. "The worst enemy of creativity is fear of the unknown. Once I worked for three years on a project that went nowhere because, as it turned out, the engineers were insulting one another in meetings. I told them to stop. Don't insult people. In two weeks we finished because they were free to ask stupid questions, and all the one-track minds dropped out. So. If you already know the answers, it's not R & D, ja? An R & D project should be very clear, but a little impossible. I always begin with the impossible. After all, what does the mind do?"

He waited, Einstein did, for my answer. I hadn't thought about it, which now seemed incredible. What did my mind do? In the window the sun was high, and the sea, at the shadowless mountain's foot, stretched sky-blue under lines of jubilant foam.

"What does it do, Matti?"

"It searches in a multidimensional space!"

Below me falls the winding stone stairwell, where, in a few moments, with dragging steps, I will leave Matti's door shut behind me. But it's okay, it's fine; the faith of my childhood has been restored with the lesson my father taught so long ago, which I feared and mistrusted. Yes, I exist in many realities. I am searching in a multidimensional space. Yes, I've recovered that faith and refitted it, and now I want to say—Dr. Einstein! This riff's for you. . . .

Okay. Imagine that you can see the smallest bits of which my body is composed, my subatomic particles. They're a sort of porridge in which space and time are blurred probabilities, not distinct places and moments. Good. Now see my atoms, which migrated into my flesh from the cores of extinct stars. The phosphorus atoms in my bone marrow are like little clocks with luminescent dials that started ticking not on my birthday, but at a date closer to the birth of the Milky Way. For all I know, they think they're still at home in the Pleiades. Now see my DNA, the ladders of generation that rise and twist and dance. Each molecule contains the genetic traces of kingdoms and phyla, stretching back through time to a single bacterium that—with barely a hint of skin to pull itself together—began, all alone, to quicken a primal ocean. This original, plain, and tiny hero is not dead. It is an ancestor present in my own genes. From DNA's perspective, space is the biosphere, from the atmosphere down to the ocean floor; and time, the four thousand million years of life on Earth. I have just described three realities, call them quantum, cosmic, and genetic: three dimensions of space and time, each belonging to different bodies that are nested, or overlap, within the system that is a human being.

Imagine Nature as a house composed of many such nested dimensions; and every mote, every body, all the bodies in Nature are its doors. There are no realities "beyond the physical"; the dimensions *are* the physical. A human being might open any door, into any dimension, through the process we call mind. In his dream, you might say, Matti opened the door into water.

Not that Matti's dream was what you'd call objectively real. Dorothy was wrong: Oz was not a place. Then if Matti's dream was not an objectively real event, what was it? What are the dimensions through which the mind searches? James Gleick, expounder of chaos theory, discusses the insight that gave the physicist Mitchell Feigenbaum his universal theory: "When inspiration came, it was in the form of a picture . . . no more, perhaps, than the visible top of a vast iceberg of mental process-ing that had taken place below the waterline of consciousness." Then, like Feigenbaum's, Matti's insight was the work of his unconscious, based on his scientific knowledge. Right, and now that I've trotted out that glib formula—the unconscious—doesn't it seem like explaining a puzzle with a riddle?

That's what I asked a physicist friend over lunch. I went off on a whole spiel. I talked about Freud's observation that the unconscious mind disregards both space and time; in a dream, things happen anywhere, anytime. And in the dreamlike state that people enter while writing poetry, pondering experiments, or just contemplating the rainbow in a cup of coffee (I pointed in my cup, got wet), it's known that brain waves travel in a changed pattern, the unconscious is at play, and people lose their sense of time. Time is suspended, you know. My father wrote about this long ago, and now there's a new line of thought, called coincidence theory, which says that the unconscious mind might actually interact with quantum reality, where normal space and time don't exist. What do you think?

"Garbage," said my friend. He didn't even stop chewing his Caesar salad. "In physics, if we say that something is the effect of a cause, we look for the force that causes the effect. If you're going to say that the unconscious mind has some sort of an effect on quantum reality"—pausing, he looked at his fork and caused a shrimp, impaled upon it, to disappear—"I want to know, where's the force?"

I digested this, but persisted. Surely, the dreamlike state is a gift of evolution, like sleep or some reflexes. What would it be good for? What if it were our gift for entering other realities? Physically entering them. Might the unconscious mind be the threshold suspended at the verge of ourselves, through which the doors of Nature open in kaleidoscopic multiplicity? I mean, I needn't be grossly dislocated (abducted by Martians!) for my body to experience another reality. A slightly altered brain function will do.

"The Yellow Brick Road is a neural pathway," I said. Just to yank his chain. My friend laughed.

But you, who have been reading this—you're shaking your head. These "realities" or "dimensions," quantum, cosmic, genetic—you object—they exist only in the author's mind.

Mind is the point! Look, my chin is resting on my left fist. My right elbow draws near, then away from my side, as the chrome pen scuttles across the paper. A depiction of the act of writing is being written, by this body, now. Is this event mental or physical? Is it happening in mind or in matter? Well? Wily reader that you are, chewing your beard, or beardless lip, you refuse this bait. You protest that it's one thing to depict a writer at her desk, quite another to paint images based on the vocabulary and metaphors of science. Porridges made of particles, dancing ladders of DNA, those are metaphors. Those are nothing but pretty pictures in the author's head.

Well, I lied to you. I said I was at my desk. Not so. Some woman called the Narrator is at my desk; as for me, I'm on eBay spending my royalties. Then whose head, at this moment, are all my pretty pictures in, including the ones of myself? That's right, they're in *your* head. Which is by way of saying, gentle reader, that such realities as we know are what we invent together using language. "To be human is to exist in language . . . together in language we bring forth our world," if you want to believe Fritjof Capra or many other thinkers who say roughly the same thing. The emphasis is on "together." Matti's vision, like any scientist's, was a working model of a world that no one can know in an absolute sense: "[T]he brain does not own any direct copies of stuff in the world. There is no library of forms and ideas against which to compare the images of perception," states Gleick. If Matti brought forth a world in which he personally vouched for the behavior of water, it was because he had a language—fluid dynamics—in which to dream, and to share his dream. Vision cannot exist unless it exists to be shared. There is no Platonic library, inhuman and immutable, wherein to end our search. What there is, perhaps, is Telling. . . .

"What are you planning to do with these stories of ours, when you get back to the States?" Matti asked. I stood in his doorway, about to go, adjusting the handbag so it wouldn't slam against my hip.

"Well, I had planned to write a long poem using the stories, but it just isn't jelling. I don't know! I need to do something, but honestly, I don't know even where to start."

He didn't seem disappointed. He said mirthfully, before shutting the door, "It's an R & D project!"

Sixteen

WHENEVER I ASKED Hemmed what their most important contribution had been, they said unanimously, "Rafael." During the '48 war, Hemmed was helpful, but it was other factors, such as smuggled fighter planes, that turned the tide. Hemmed's importance lay in having turned a gaggle of amateur inventors into the institutions that would develop Israel's innovative technologies for decades to come. I never expected to visit Rafael, but someone swung it for me.

As soon as I got permission, I pictured floodlit, scaffolded towers in which rockets hung suspended, looming across a dust-blown altiplano. American thinking.

Entering the front office, I was confronted by a young security guard whose eyes focused on mine so exclusively that I panicked and blanked the name of my contact, let alone the other codes, and fumbled in my handbag for Post-it notes. There went my fantasy career in the Mossad.

"Good," said the guard at last, "use this phone."

Inside the office was a crowd of youngsters under army age; their brown ears twinkled with silver, and one notion chased another across their constantly turning faces. They made a great intimate racket, chattering and arguing, pressing toward the counter where a heavy-jawed clerk fetched ID badges clipped to a board. Morning sun glittered on the girls' bare shoulders, and turned the boys' curls to molasses candy. How pretty they were. How old I felt! Later, I learned that the children of Rafael employees are allowed to visit and hold small jobs. I possessed a similar security clearance in my genes: the Museum of Early Weaponry, while off limits to the public, may be toured by descendants of Hemmed.

A turnstile stood at the office exit. Through it came an old man who raised a long, splayed nose in my direction and smiled, with small yellow teeth. His name was Pesah. We shook hands, and immediately we had a problem: the counter clerk insisted that no ID had been issued for a Bentov.

"I had Tzippi arrange it," my host pooh-poohed. The clerk made some phone calls until, satisfied, he handed over the receiver. Pesah spoke into it.

"Tzippora," he said with relish, "why do you complicate my life by never doing correctly what I ask the first time?" We left the office, and after a short stroll entered a building that could have been anywhere in Israel: plastered corridors, wooden doors, by no means new.

"To your left, please, and welcome to our museum."

It was no more than a linoleum-tiled hall some forty paces long, yet it had the authentic restfulness of museums. By the door hung a framed photograph, enlarged almost to the size of the man it depicted. He wore a military beret, and his magnetic, skewed face gave me a jolt.

"Genka Ratner," I said, backing off.

"That's the Jew." Pesah did not yet escort me into the hall. His tour began in his office, which was furnished with three schoolroom chairs, a desk, and a deal table on which hummed a

teakettle on a hot plate. I made a cup of Wissotsky tea. Pesah
stooped under the table to offer, with a flourish, a wastebasket
to receive my used tea bag. He apologized for the quality of
the brew.

"I like it, it reminds me of my grandmother. She used to call
me 'little Wissotsky-chai.'"

"Russian of course."

"Italian?" I asked, but Pesah was Romanian. "I thought,
because you're so polite."

My host looked grave, evidently flattered. He said that we
would now listen to the news report, as a bus had been bombed
this morning in Tel Aviv, resulting in twenty casualties, and no
doubt the prime minister would speak. On his desk was a wood-
veneer box that I'd noticed after glancing about for a TV or
radio. As soon as Pesah dialed an in-house number, the box, an
archaic speaker, filled the office with the lowest and coldest voice
of Yitzhak Rabin.

"We will keep a cool spirit." Each syllable was rock crystal. In
the background rose an outcry of yells; Rabin was at the site of
the bombing, and sirens wrapped around his words. Where he
was, was confusion, blood, and madness. "We will keep a cool
spirit," he repeated. "We will not allow our enemies' acts to
deflect us from our goal." (On the day of Rabin's assassination, I
would call Sara Gil from the United States; and she would say to
me, "They have shot the heart of my generation.")

Like Pesah, I remained standing while the prime minister
spoke. There was one pretty thing in the room, I observed: the
window looking into an oleander's leaves, like emerald sickles.

Pesah switched off the radio and we made ourselves comfortable,
amid rueful comments. Then my host said something wonderful.

"Now, I have a surprise for you. But tell me, is it true that your father never mentioned us? Really never spoke of Hemmed?"

"Not to me." The curator absorbed this without comment.

"We acquire most of our exhibit," he continued, "by accident and hearsay, from junkyards, from under beds, from people's file cabinets where a memento sits and rusts, from the corners of courtyards. Recently, we got an old Tolar that had been lying beside a building in Haifa, in a field."

"A Loretta!" I was half out of my chair.

"Only in Hemmed were these recoilless cannons called Loretta. The correct term is Tolar. Its provenance is not certain, but it is dated 1949, and to all appearances"—he showed his little corn teeth—"it is the work of the hands of your *abba*."

It crossed my mind, as Pesah and I walked into the gallery, that my excitement was that of a little girl who used to climb down the laboratory stairs to find her father. We did not have far to go. An area had been cleared around a large contraption like a plunger mounted on a tripod, suction cup tilted toward the floor. It was painted rust-red. I came closer and looked at the five-foot cannon barrel whose breech end was like—I was reduced to domestic similes—an iron flowerpot. I knelt on the floor to inspect it. In the flowerpot's base were two curved slots.

"Gas vents," I said, in the direction of Pesah's scuffed sandals.

"You really understand!" I looked up and laughed.

"I'd be stupid if I didn't—everyone in Hemmed talks about it. And this round hole here?" The curator stepped in closer, bending cautiously from the hips.

"That hole, yes, that is where your father, of blessed memory, would have used an electric spark to ignite a black powder fuse, and the cordite that was packed in this end."

He shuffled back, watching in respectful silence as I sat on my heels. My skirt pooled around my feet. I wondered what I was supposed to feel or do.

I recalled what Shai had written in *Hemmed Days* about the testing of the Loretta. Various generals had gathered on the beach, where Tel Aviv's luxury hotels stand now. There was the sand, the sea, the setting sun, and a cannon barrel suspended horizontally on a rope. If it worked, they'd have an antitank weapon. The watching crowd was as silent as I was now. Ben-Gurion was in that crowd. For months he had bullied Hemmed: We need bazookas, now. We need flamethrowers, now. You have Jewish brains, don't you? What we don't have is time.

A physics student approached the suspended cannon from the breech end and ignited it. The blast from the vents blew a pair of pliers out of his hand. The shell sailed out. The cannon hung level on its rope, not swinging, as stable as if nothing had happened. Did they cheer? No. They talked rapidly in low tones as they beheld the cannon barrel doing nothing. An action, a shot, had taken place without an equal and opposite reaction. To those whose fate hung in the balance, perhaps it seemed as though Nature had agreed—in their necessity—to wink, and allow herself a forgetful moment.

I thought, My father made this weapon with his own hands, like his cardiac catheter. The Loretta was designed to cheat the law of action and reaction, to pretend that an event did not occur. The Loretta erases the past—it is like a symbol of forgetting. Something about this is important.

Besides that hunch, I felt . . . well, that a feeling must exist for this situation, but it hadn't been invented yet—only the raw material was there. I used to feel this way in the corner bookstore when I was young, shopping for Father's Day cards, pulling them one after another from the racks, looking at pictures of golfing dads, fishing dads, baseball dads, dads in slippers with the newspaper; and leaving empty-handed. I fingered a little rust-red key that dangled on a chain from the Loretta's C-bracket; Pesah explained that it was for azimuth adjustment. For an instant it was everything I craved. I rose, dusted my hands, and asked Pesah if his collection included a Hollow Charge. Maybe, after all, I would find the elusive brass shell I had wanted so much.

There were a few score shells lying in rows on metal painter's shelves, the best cases that this museum could afford. The shells were finned and finless, perforated and solid, blue and gray; my host and I rummaged among them, handling and pushing them aside, like shoppers at a flea market. Confident that we would put our hands on a Hollow Charge, after a little while Pesah was puzzled and bemused.

"We made thousands. . . . I remember the girls cooking the TNT in a double boiler, until it was soft like nougat. It smelled of almonds . . . we must have one somewhere, it's not possible . . . and they used a winch to pour it out. It's not possible," he repeated, lightly touching the shells, then looking at his hands, as if they too might vanish.

I should have known. To take him off the hook, I asked about the Automatic Dog.

"You're standing beside it," he replied, pointing.

"That?"

For months, I had tried to wring a description of the Automatic Dog out of Hemmed. I'd gotten a torrent—a torment!—of unhelpful negatives, ranging from "not a dog" and "not very dog-like" to "not what you think." Now I crouched over the artifact, putting two and two together. Finally, I shook my head.

"It, ah . . . sure isn't like a dog."

Pesah shrugged.

"This has certainly been a delightful tour," I said as we wound up in the doorway beside the framed portrait of Genka Ratner. "Do you ever let history teachers bring their classes here?"

"It is my dream," confessed the curator. "Now when youngsters have the most sophisticated technology even as their toys, these weapons appear—funny. It seems like nostalgia, but it isn't nostalgia, to say that we rescued ourselves with what we made from nothing. This was real."

"Yes," I remarked, "and it wasn't funny, either."

"It wasn't," he agreed. We walked together from the museum back to the gate office. A guard came over to buzz me through the turnstile, but Pesah waved him off.

"I wish to say something more, if you will allow me," he pronounced, lowering his nose but fixing his small eyes upward, hands clasped behind his back. "Something personal."

I listened carefully while he spoke. When he'd finished, we shook hands again; he promised to send me a snapshot of the Loretta, if the authorities could be persuaded.

Israel is a half-day country. It takes some six hours to drive its length, from the Lebanese border to the Egyptian. Having ridden these roads many times, I knew their familar rhythm as the bus rumbled south. Past the purple sea washing the medieval pier, where frogmen used to prac.. e-dive. Past block towns with billboards staggered at their edges, and the yellow gas pumps of Paz. Past fringes of prickly pear where an Arab village once stood. Fields passing, again and again, iron-red, dull green, or flaunting arches of bright spray, between cypress windbreaks.

The Mihmoret Pancake House, a cowboy lariat on its billboard. The huge spooky pipes and tanks of Nesher Brewery, run by the Tempo firm, whose name made me imagine bottled time. And always accompanying us, the steel cruciforms of the power towers, their swooping cables strung with colored balls, like whole notes on a country-long musical stave.

In a certain spot: symbols on the road that had nothing to do with traffic. Or so we all had heard. There, under the tires of the country's buses, trucks, Jeeps, Peugeots, Subarus, and Mitsubishis, pummeled by the wheels of daily life, are the runway marks for the final airlift. Should all the doors of Israel be aflame, there was the final exit: into air.

No one spoke aboard the bus except the prime minister. This was a scene I'd lived before. I smelled the mildew in the bus ventilation; I looked through the grimy oblong window at the passing coast. Breakers slid to shore and broke, foam-crusted, in a repetitious shimmy. A standing passenger, balanced in the aisle, would have seen all the windows connected by a series of glittering blue-and-white hyphens. But no one stood. Rows of heads joggled in the seats before me. Israelis are loud. Tell it to the hundreds of buses, porting hundreds of thousands, perhaps a million, gently shaking passengers, who sit in silence while the bus driver turns up the volume of the news. This is the procedure that follows terror in Israel, when the prime minister tries to build courage back inside each citizen. Word by word.

But this time, I wasn't listening to the prime minister. I was mulling over what Pesah had told me at the turnstile, as I was leaving.

"If you'll permit me to say this, I understand your father," he'd said. "Why he didn't speak of certain things. There is a whole generation of us like that. We ran away from a burning

Europe, and coming here, we knew there was nowhere left to go. Your father was how old when he came here? Seventeen? Then—remember, Hitler came to power in 1933—he would have been a little boy. He would have known that the state wanted his blood. He would have lived, as I did, with that fear. I remember perfectly well what it was like to grow up under fascism. You have to understand the fear he grew up with."

Seventeen

M Y FATHER DIED during the lilac season in Boston, when the grimy snow that has gripped the streets all winter seems to return, after a muddy month of thaw, as the fresh white clusters it would have liked to be. I was twenty-two, working in a bookstore and living with my fiancé, an engineering student named Adam. We rented an attic apartment above an old Armenian couple who were our first landlords. Everything, including love, was new; and at the time, I remember, I had a private image of my life. On the day we moved in together, our bed was hoisted up to the dormer window, where I leaned out alongside the pulley ropes. Down in the yard, Adam was shouting directions; he was a lanky, high-shouldered boy who shouted through soft red lips and a growth of virgin beard. Our new pink-and-gold mattress hovered above the topmost leaves of a horse chestnut, and that image was the most beguiling I had ever seen.

When I remember Adam, I think of the odd tactile illusion when we held hands. His broad fingers always felt as if two of them were laced between each one of mine.

It was Adam who broke the news of my father's death, one morning, after his parents called us because they'd just seen the newspaper. But never mind that. A timid person can avoid newspapers—can try to dodge the past, as well as duck the future—so it was years until I mustered the determination to read that issue of *The New York Times,* by then on microfiche, to confront the details. I copied off the relevant pages, and filed them away for another decade in one of my suitcases. Now I'm gingerly finger- ing the old paper, with its slick unwholesome texture and hue— grayer, I think, with the years—looking for quotable headlines.

THEY TALKED OF THE TERRIBLE ROLLING THUNDER
WORST U.S. AIR CRASH KILLS 272
FAA CHIEF WEIGHS GROUNDING OF ALL DC-10S
FLATTENED DEBRIS AND "BODIES ALL OVER"

I don't particularly care to remember any of this; indeed, the only reason I continue this irksome story—well, as it happens, I just forgot that trenchant reason I'd memorized, the way it is after passing your fortieth birthday, when you're smack in the middle of the room you just entered, looking about yourself for—glasses? Tylenol? car keys? cup?—and no one is home to remind you, you citizen of the great state of Never-Never Mind, frontier of Death, entitled, as you now are, to learn that when the human yearnings are all lined up, at their head stands the yearning for the names of things, the right and real names.

Where was I . . .

The sunlight of a bright spring day, turning the hush of a fortieth-floor office into something almost churchly. I had an appointment with someone I'll call Lou Grimmer, the senior partner of Grimmer & Grimmer, and his junior partner, Frank.

·   ·   ·

Normally, such business would have been handled by Irv, my father's estate attorney. But Irv had fallen out with Sonia since he had executed my father's will, which bequeathed a small sum to my mother and almost half to me. Sonia could not fire the executor, but she hired her own private counsel to dispute him on such points as the yard gold. My father had buried gold coins in his front yard, "For those times," Irv chuckled, "when the government falls and the banks close and there are tanks in the streets. Your father wasn't the only client I've had like that." The will gave Sonia all household objects. Gold coins, claimed her attorney, were objects. Buried, in effect, in Sonia's household, so Sharona should not get any, bibble babble bubble. Whenever Sonia's lawyer contacted Irv, Irv contacted me, in a foul humor. "I haven't come across the name of this Buck Morley whom your father's widow has seen fit to retain. I can't find him in *Who's Who*. He didn't go to Harvard." Sonia visited Irv and screamed. No one in his forty years of law practice had behaved in that manner to J. Irving Waldstein, Esquire; she actually screamed, like a—peahen! He threatened to resign as executor, which would have left Sonia in charge of my inheritance. I told him over and over how sorry I was, how grateful for his patience, and how sorry, and how grateful. My father was not a month dead.

Irv could not represent me in these disputes with Sonia because he was the lawyer for the whole estate, and all its heirs. Conflict of interest kept him from representing me alone. Consequently, he had me hire some outside lawyers, to be remunerated from my eventual inheritance. Their fees made me anxious; I was earning minimum wage at the cash register in the Crystal Phoenix New Age Bookshop, a job gotten not so much on the strength of my Princeton degree as from my connection to the famous author Ben Bentov, which I revealed in my interview with the pimply shrimp of a manager, who wore charms against demons. This was the sole employment I could secure through my father's influence, for reasons I will tell soon.

As Cicero says, *quae cum ita sint*—these things being as they were—when Sonia brought a suit for wrongful death against Alacrity Airlines, she did not allow Irv to represent her, which would have been the usual thing for the estate attorney to do. Irv called on my haunted telephone—the same one that had transmitted the news of the crash, then a many-voiced babble of legalese.

"Sharona. Are you aware of what is meant by the term 'ambulance chaser'?" Sonia (he huffed and puffed) had hired Grimmer & Grimmer, Inc., to sue Alacrity. "That woman hasn't included you as a plaintiff against the airline. Grimmer didn't know until I told him, 'Bentov had more than one heir.' There isn't a soul to speak for you. I can't do it. You must go down there and talk to them yourself."

Lou Grimmer's headquarters were in L.A. His Boston office was borrowed, the desk cleared around a speakerphone and an onyx ashtray the color of pork fat. I came in, shook hands, sat in a Windsor chair. Grimmer hit a speakerphone button and asked, intriguingly, "Can we get a girl in here?"

A secretary sidled in. Grimmer folded his arms on the desk, bulged forward, and looked me over. My jeans were torn, my hair cropped very short, and I wore a T-shirt. Blazoned across my chest was the image of a bare heel in a flying roundhouse kick, and the emblem of Kim's Tae Kwon Do Academy. I was staring at him in perfect cluelessness. His shirt was beautiful. His big supple cheeks had a bloom, as if just splashed with cologne, while his eyes were elderly triangles. He glinted at me, curious, polite.

"You realize—you use your father's last name? Miss Bentov?—that you come as a big surprise to us."

"Really big," chorused Frank beside his boss. He was young, and had freckles.

"Until Irv called us last week," Grimmer said in his rich, flexible voice, "we didn't think Mr. Bentov had any children."

"I can explain. My father's wife—"

"Sonia," Grimmer reproved.

"Sonia," I apologized. They were on a first-name basis, then. My gaze shifted to the monochrome shelves of statute books. It struck me that if someone had treated you strangely, you would sound strange yourself if you described it. You would sound all wrong.

Lou Grimmer smiled. His smile was like a lock.

"Since you do come as such a surprise, Miss Bentov, I'd like to start off by asking you a few questions. Did your father smoke?"

I laughed. Then I realized he was waiting for an answer. I said he did not.

"Did your father drink?"

Nervously I said no.

"What was your father's religion? What church did he attend? . . . You're not sure?"

I began to mumble that my father was Jewish, but in this country he had become a follower of Mahananda, like the Beatles, and he had written a book. Grimmer began wiping a pair of Gucci bifocals and murmured an aside to Frank.

"You see, Miss Bentov," said Frank, tapping a pen on his clipboard, "when a man in your father's financial circumstances passes away, there's always a question when relatives start turning up unexpectedly. Not that you crawled out of the woodwork, or anything. We know that, because your father's will does leave you—"

"More than I'd leave the child of a divorce!" Grimmer chuckled. "Maybe your dad felt guilty. No offense." Calmly, he touched a gold lighter to the end of a cigar. I had never seen a rich, fat man light a cigar except in old movies. Wow. Wait till Adam heard.

"Your father's widow insists that you had no relationship with him. She—I'm putting it in her words, now. She says you didn't exist for him."

"She says I don't exist?"

"For your father. That you had no relationship with him."

My heart went white. It felt like.

"Uh—I can explain. Thing is . . . he had to pretend to his, their, 'social circle,' I guess, ah, that he didn't have a child. He wasn't supposed to mention my name and, also, he couldn't see me after 6:30 p.m." I refilled my lungs. Only their tops worked. "So we were always alone together. It was just me and my father, and that was our—it was okay. I never saw his friends until the memorial service, and, um, they, ah, hadn't heard of me." It came out goofy and breathless. Grimmer relaxed, leaning backward in his creaking leather chair.

"I can understand that. A second marriage. Wouldn't you say that's understandable, Frank?"

"The child of a divorce, sure."

A divorce had given birth to me.

Grimmer exhaled. Smoke drifted from his insides, slowly, sort of gray paisley. The air turned odorous.

"May I call you Sharona? Sharona. What was it you called your dad, by the way—Daddy? Father?—okay. Let me give you a

219

little of my philosophy about this lawsuit. I think that what makes a lawsuit big, and we are not talking about sprained backs and herniated disks. We are talking about a very serious business, and I think what makes cases really substantial, probably more than anything else, is fundamental honesty that rings through the whole story. If juries say, 'This is absolutely right,' if it's a true story and it's absolutely solid, then you get the maximum. You don't have to build up this case. We want to explain the losses for what they really are."

We exchanged our first look of mutual comprehension. A film of oily smoke coated my face, and I kept from licking my lips. Frank's pen was busy scritching.

"And if we can do that," Lou suggested, "if we can demonstrate the sensitivity of these losses—not the dollars and cents, because if that's what this was about, the recovery would be very small. We're discussing the loss of a unique relationship, of a father who would have advised you, and stood by you, and given you the benefit of his character or attributes, presumably for a good many more years—if we can do that, then we get a lot of money. Makes sense? Okay. So if we proceed with this case, what we need from you is some, or a good deal of, evidence of your relationship."

"Evidence . . ."

Lou spelled it out.

"He must have given you a pearl necklace." He tipped the chair back as if to see around me, where the evidence might lie.

"I, I don't have one."

"One of those add-a-pearl necklaces, when you were a little girl? Everybody has one of those."

"My father gave me an optical microscope," I said on reflection.

"That's just a toy. You're sure, no pearl necklace? The first I've heard of a man of his . . . what did he give you, then?"

"He gave Adam and me a clock when we moved in."

"Do you know its value?" I described it. "Oh, I thought you meant some heirloom clock. What about a gold watch, a charm bracelet? Things a jury can understand. Things a real daughter owns."

I needed a pearl necklace to be a real daughter! That was why I'd never felt like one! No one had told me before! Cool, milky, hard round gems, in a row, in my fists. I wanted one. Right now. Right now.

"Mr. Grimmer, he didn't buy jewelry. He said it was bourgeois, and none of the women on his kibbutz wore it."

Both attorneys grimaced, and consulted each other about how Bentov's communist background would play; maybe Jewish jurors could relate to it.

"He sent me to Princeton?"

"Go, Tiger!" cheered Frank.

"Thank you." Mouths did go dry, how about that.

"Ivy League schools," Lou demurred, "are not necessarily the high road to a jury's heart. Let's approach it this way. Forget the gifts. What are the things that, in your mind, prove to you that your father loved you?"

If my story was true and absolutely solid, then some proof had to exist that my father loved me, or why would I have believed it for twenty-two years? Without a pearl necklace. I hung my head. On the onyx ashtray rested an inch of ash, wrapped in a veined leaf that had carbonized with its pale tracery intact. One jarring tap, and this last remainder of the living leaf was gone. Engrossed in the sight, I waited. I was waiting for my father. He would stride into the office and deal with these uncreative types in short order. He would wrinkle his nose at me. He would laugh away the evil spell that allowed strangers to say whom he loved, or didn't.

When I raised my head, I saw Lou Grimmer. Behind him, the brown rankle of skyscrapers that was Boston, stretching toward the copper-blue Atlantic. The view shifted slightly; the tower we sat in was buffeted by wind. Everything I saw was real. This was it, reality. This was the only reality I lived in. I had been ignorant of the rules that made things and people real, like a real daughter having to own a pearl necklace.

"Well . . . he called me nicknames."

They asked what sort of nicknames.

"Pookli . . . it doesn't mean anything. It sounds like a Hebrew word for belly button."

"Aw, that's cute! Lou—it's cute, and it reflects the unconventional kind of guy her father was. Say, did Sonia keep him from writing you? Any little fatherly notes?"

"He sent me, once, a valentine he drew himself."

"There ya go!" Frank's pleasure was infectious. He was a nice man.

They were both, in retrospect, good men. They were keen, adroit, and wise in the way of master craftsmen. I wore a pink ruffled blouse and suit to the trial, several years later. We lost. Now how about this. Reviewing my faded files, I see that those three questions—smoke, drink, religion, impaled in my memory—were posed not by Lou Grimmer, after all, but by an aggressive lawyer for the airline's insurance company, during a deposition. There were too many lawyers, too many similar scenes; yet I marvel at my erroneous memory. For so long it had the sting of truth. I would have sworn to it in court: it was "the best of my recollection." In what unholy way does a false memory become your best?

But my valentine! Here it is.

He must have had the card lying around. I wonder. On the front, a stylized woodcut of Jerusalem: wrought-iron windows, cats, cypresses, domes. Inside is his hand-drawn cartoon. A bearded old man, all shaky lines, bent over a cane and trailed by the words "COUGH, COUGH, COUGH. . . ." Above him is a big heart, Magic Marker red. Through a hole in the heart shoots, with propulsive force, a projectile weapon whose breech end looks not so much fletched as finned; it is labeled ZAP! A banner, winding all around Mr. Cough Cough and his levitating heart, bears the greeting TO MY BEAUTIFUL POOKLI FROM YOUR OLD MAN.

I remember how, photocopied for Grimmer & Grimmer, the red heart came out black. Something parallel began happening to me around that time, because I remember how unkind I was to Adam. After a fight, shortly before I broke our engagement, I came home to find him leaning out the attic door, into the stairwell. He couldn't wait for me to climb the stairs. Instead, from above, he held down to me a spray of purple Japanese irises, and their slim stalks were trembling. I saw this, I remember, with scornful disbelief.

Eighteen

ONE YEAR AFTER my father died, I was sitting on the floor with the telephone in my lap, dialing his number. I had moved to another attic, in an eighteenth-century house. The oak planks under my crossed legs were thicker than doors; the forged nails, headless and deep in the wood, gave me the casual amusement of looking for them. When Sonia answered, I swallowed and asked her about an item in my father's estate tax return, labeled "Funeral Expenses." There had been no funeral. Because there had been no remains.

"The airline sent me a body bag," Sonia replied, not pleased.

"But . . ."

"It was very hard!" Sonia controlled her lisp, which sounded so hostile, and continued, "It is interesting that you call today, because I am now arranging to install the urn in the columbarium."

"The what?"

"The urn. You may go there and meet me if you want. It is in Newton."

"What, what's in Newton?"

"The columbarium! Where the urn is stored!" I hunched

down closer to the phone, a smooth pyramid that weighed about four pounds and had plastic horns, bells, solidity.

"But why didn't you tell me there were remains?"

"It was very hard," she repeated, slowly, for my benefit.

"But he was my father." My fingers were tracing the wood grain in the floorboards. Under my fingertips passed the contours of mountain ranges, vales, and frozen seas. A world mourned inside the dead wood.

"I know that I did what was right."

"But he was my flesh and blood!" Blurting this cliché, my tongue felt like a four-inch blob that could not have spoken for me all my life.

"Oh, you don't understand," Sonia drawled, rather loudly, as if someone, some audience, were at her side and listening to her, there in my father's house.

The urn installation was the third opportunity I'd had that year to meet my father's widow. The first time had been after the memorial service, in his, their, house. People lined the carpeted stairs to the upper level, gripping the banister behind their backs, talking all at once. As I sidled past, a few gave me friendly looks, and others, unfriendly. Sonia in her gray cashmere darted back and forth, arms held out, like a little screech owl scared into the daylight, driven to and fro.

"Take, take, it's what he would have wanted!" she kept crying. People filed into the guest bedroom; the chattering line bore me along. I saw my father's shirts hanging in a closet. All of them. All his seasons. Below them, agape, were his shoes. I kept moving. People loitered by the bedside, browsing over two cartons filled with men's cotton briefs. I thought that was unbearable. I ducked out through the press, and crossed the hall into my father's study, shutting the door. Sunlight, silence. I took from his desk the clear square containing a silvery tree.

The second time, Sonia and I met in an atrium café in Harvard Square. Lou Grimmer insisted that we negotiate the percentages we would receive from our suit against Alacrity. I remember the echoing café with its iron furniture, and Sonia's prim lowered eyelashes.

"I don't care about material things," she began, "but Buck—my lawyer—says that you should get ten percent."

Perhaps in Sonia's circle, deals began with this kind of disclaimer, much as a Japanese business letter begins with a reference to seasonal flowers. I had already heard, I would continue to hear, about my fellow plaintiff's disdain for material things, which, however, concerned her lawyers very much. It pissed me off. At that crucial meeting, though, I felt no anger. Hell, I couldn't even taste my pricey asparagus omelet. My mind wasn't on business. This was the first real conversation I'd ever had with Sonia, and I'd looked forward to it. We both had. She wanted to explain herself to me, and she did.

"When I told your father that I wanted a child, he would say, 'I already have a child.' When I said, 'But what about us, what about our family?' he said, 'I already have a family.' It wasn't fair for him to deny me a child because you were there." Sonia spoke her heart. "It wasn't personal what I felt about you." She offered this thought, which she had labored within herself to clarify. In good faith she repeated, "It was never personal."

And there we were, everything settled, lunch bought and paid for, standing on the windy corner of Brattle Street and Mt. Auburn Street. I kissed her and said, "I love you."

She smiled and replied over her shoulder, in the June breeze, as she started to cross, "Oh, I too . . ."

Now, I did not, I could not, possibly have loved Sonia: she scared and depressed me. But I didn't not love her, either. What I'd uttered could not be calibrated according to false and true. Something had switched off in normal reality, and switched on in an unknown place.

Many years afterwards, I began to send Sonia letters. I demanded explanations, quite forgetting that she had already explained herself. Or unwilling to accept her story. Or still sore, I guess, that I had said "I love you." I wrote with venom, spite, and childish rage. Sonia answered with holier-than-thou cattiness, offering to pray for my soul. In short we were a pair. At length, I wrote like a civilized person, and Sonia answered with a brief apology, which I accepted. It had dawned on me that forgiveness is sometimes a matter of emotional economy.

On the day I found out about my father's remains, Adam drove me to the Newton cemetery. He left me at its feudal gate. I meandered around the lawns and found the columbarium, in a Tudory cottage. The oak door swung smoothly inward, and stepping inside, I saw straight down the room to its end: a picture window draped in climbing plants, looking out on a . . . grassy knoll, I supposed. Air freshener perfumed the sunlit calm. Someone had designed the view to draw me into the columbarium without fear, so I walked in, across an echoing floor. Beige marble columns divided the room into aisles lined with glassed-in shelves, where the bronze urns stood in separate niches. I was reminded of a safe-deposit vault at the bank. Each niche had a brass bud vase, with a silk posy, where a deposit box had a lock.

I found Sonia at the end of an aisle, her back turned, not hearing me or anything else. I greeted her, and she answered civilly. She wore the muted tweeds my father liked; her tote bag slumped on the deep windowsill, among pots and pots of silk bouquets. It seemed she was about to remove the urn from her bag, and was contemplating the niche where it would fit. She lifted something in her hands, slid it into the niche, at waist height; then laced her fingers together and regarded it. I stood beside her. Her eye sockets looked chapped. In the niche sat, all alone, a bright brass cube. It bore a plaque with the engraved image of a stump, sprouting a leaf. Below, my father's name, his dates, and the Hebrew phrase ETERNAL LIFE.

"That's beautiful," I ventured. "That's really beautiful." I thought it looked like a Disney cartoon.

"Thank you . . ." Sonia fetched a profound sigh. She rubbed her rough sculptor's knuckles in her eyes. "I did what was right." Reaching into her canvas bag, she brought out a long object and a packet bound with a rubber band. "You might want these," she offered, looking in my face, in a way that classed me among her former problems.

Outside, I could not wait but went greedily to a bench by the exit and laid my heirlooms in my lap. The packet held photographs of me at various ages, taken by my mother, and kept, surprisingly, all these years by my father. The other object, unwrapped from its brown paper, turned out to be a photograph album.

The July sun was crisping my arms, but I turned the pages slowly, fascinated. Here was my father inside a sunny tent. He bent over a drafting table, holding a paper by the corner, smiling; Matti sat beside him. Here he was in the middle of a village street, swinging a wine bottle by the neck, his arm through Aia's arm, her hair floating, her arm around Sara Gil, demurely hand in hand with Uri Gil, all of them trailed by a donkey. Here my father sat in bathing trunks in the sand, while Plafchik lounged on a canvas chair. Here he and Misha, in army jackets, posed before a low building with narrow windows. Over the doorway: WEIZMANN INSTITUTE OF SCIENCE. From their four boots stretched a single shadow, a thin wing across sandy ground. Here he was in a white Russian blouse, arms thrown wide, pitching exaggerated woo to Ziva, her face scrunched with laughter.

I didn't know who those people were.

I guessed who the boy was, among the young workers in sun hats, sorting cauliflowers. Then I gasped at the ID photo from the immigrant youth village. He was glaring: my boy father, miserable with rage. Though in the photograph I lingered over longest, he seemed pleased enough. His hair was parted neatly on the right side, the opposite of the left-hand part on the smaller, less rascally boy beside him. The woman had clenched lips and sickles under her eyes, radiating tension. The man had the stern brow of an impartial judge. He wore a tie, vest, jacket, silk and wool; I could see the quality. Closest to me—had I been there—sat a chunky little girl in a plaid dress, arms crossed like a tomboy, and on her face a most delightful grin. Of utter confidence.

His little sister.

Auntie.

Nineteen

WHO GETS the last word? Second row, medium height, and rather comely. He's looking sideways at the woman craning out from behind him, bent from her hips in some jovial gesture of upstagement. I don't know who she is. Shai does . . . but at the moment of the camera shutter's opening, he seems to contemplate a phenomenon at his side, not a woman. His expression is absorbed.

Beside me at the Gils' candle-lit table, Shai wore a starched shirt and silk suspenders; he rested his elbows on the table and laid his chin on his crossed cuffs, breathing like an old dog. Then he hunched forward, chin nearly grazing the linen tablecloth, and forced his shaking hands to manipulate cutlery. He slowly sawed a bit of glistening chicken topped with half a limp apricot. He herded, poked, and levered the morsel into his mouth, his face expressing nothing but the material strain of its component muscles. He rested a few moments, chin on cuffs again, before the lesser effort of chewing and swallowing. I watched from the corner of my eye, since no one would have dared observe him directly except his wife, a warm-voiced brunette with a con- trolled manner. All through dinner, this lady monitored her husband with a flickering regard that included me; and I replied, in the same silent dialect, that I understood the need for care. After satisfying not his hunger, surely, but his dignity, Shai abandoned his chicken cutlet, his diced salad in the cut-glass bowl, his pink wine in the crystal cup, and all the other torture implements of Sara's well-laid table. He leaned back; head drooping to the side, like a train passenger falling asleep. It was a better position for conversation. In a thin steely voice, he told me that his area was particle physics, and that he studied a particle called the B meson.

"B stands for Beautiful," he said. "There is . . . another meson . . . called T, for Truth."

"Oh, Keats," I said, and a benign judgment moved across his pallid face. I asked why he studied the beautiful meson. When the universe was born, he replied, in the big bang, there should have been equal amounts of matter and antimatter. Now, however, we detect far more matter. Therefore the universe is lopsided. He wanted to know why. I was touched that a man who'd been given his powers of mind and character, and then polio, should question the balance of the cosmos. Since he was not in a condition to talk freely, I thought carefully about what to ask him.

"Do you have any stories about my father?"

He pondered this drowsily, with round-lidded eyes. From the start, he said, it had been clear that my father would leave Israel for a place more suited to his ambitions.

"Not just ambitions," Uri interjected with a hurt look, and I recalled that he and my father had been close.

"Not ambitions, exactly," Shai qualified, "but you could see it on him, he was so curious. The . . . inventiveness that filled him . . . to his fingertips. Itzik lived for technology. It is therefore ironic . . . that technology killed him."

There was a pause to which everyone seemed accustomed, as if it were generally accepted that Shai said this kind of thing. I didn't find him touching anymore. He was scary, summing up life and death like that. I might as well ask Beauty, or Truth, to tell me stories. Instead, he'd given me a cold, cold epitaph.

A few weeks later, he gave me an epitaph I liked much better—it could be my own.

It was at the end of August. By now, I regretted having to leave Jerusalem soon. I enjoyed doing laundry at #4 Arlozorov. I had a washing machine in the bathroom and a folding rack on the balcony. In the afternoon, I would stagger from the washing machine, arms weighted with a ball of soggy clothes, and squeeze myself past the balcony door, apologizing to the grapevine tangled in the wrought iron. Within the garden shaded by tall hedges, the balcony, tiles ablaze, was like an altar consecrated to the sun. As I set up the rack, the stone wall behind me baked. I had to wear sunglasses or I would have burned my retinas. The strung clothes instantly began fading. A geranium-red linen shirt would glow like a ruby, showing the irregular grain of its woven threads, and in minutes turn matte pink, far and away lighter, as though disburdened of some passion. Instead of hanging drenched, it would begin to flutter a little in the breeze. I wore a straw hat whose shadow crosshatched everything I touched. Once the laundry was hung, I would sit against the wall, knees drawn up, while the right instruments would have detected a comet's trail of moisture streaming out of my body toward the heavens. I would think about the people I had met, and what I'd learned. About Hemmed's ant bomb. Or the general who stole prototypes he needed to use that afternoon. And the chemist who'd told me how technology usually developed from centuries of tradition, but Hemmed was "a mutant that mushroomed overnight, because of the instant feedback we got from the battlefield, and because we had no traditions, no rules to break, so any crazy idea was tried. Instead of tradition, we had improvisation." And the engineer who'd opened Israel's first oil well without the proper equipment for it, using, he'd joked, "Hemmed-like improvisations." I thought about a model Loretta, I was pretty sure it was, on a shelf in Ben-Gurion House, that I'd shown to a bored, mistrustful young docent. When I reentered the kitchen it seemed dark and cold, and I would ache with relief, drinking glasses of water that I could feel replenishing my flesh back to the kidneys.

On Friday afternoon through Saturday night, I would visit with friends. How enjoyable to dress up, penciling a green line along the roots of quivering eyelashes, in the mirror, while a sweet tenor voice drifted from an apartment across the street, practicing scales and phrases for the Sabbath service. On solitary nights, I watched TV—once, a live broadcast of a drama class taught by a celebrated actress of the Tashah generation. The play was the classic tragedy *He Walked in the Fields*. They were rehearsing the scene in which a young kibbutznik is recruited by a Palmah commando.

"Frankly, now, if within the bounds of censorship," the actress asked the boy playing Gingi, the commando, "what did you feel upon receiving your call-up for army duty?"

Supple in an undershirt and sweats, the actor looked his teacher in the face. The camera circled the two artists, whose bond, at that moment, was itself a powerful study.

"I had a serious debate with myself whether or not to join a combat corps," he answered.

"Well, you see, none of that existed in your character's time."

The actress broke away from her pupil with a swift driven turn, and addressed the rest of the repertory group lounging in athletic postures, attentive and puzzled.

"This business, at which we're all now international experts, of 'How does it make me feel' and 'What does it do for me,' this 'But how do I feel, *oo-ee!*'—it wasn't there yet. There was no *oo-ee!* Reality was different, and naturalness was different. You're all trained at understatement because it looks natural. We were not so self-conscious. There wasn't time. In our time, overstatement was natural. Not of one's deep feelings, which one hid, no, but the feelings of hevrei, the group feelings, were even exaggerated. It was a time of large gestures and declarations. This is Gingi, eh?"

She threw out her chest, swaggered over, and swung her arm smack into her startled pupil's back, while the rest laughed. "Now, when Gingi starts talking to Uri about the platoon, that's real. Then his gestures vanish and it all comes from the heart, because that's the Land, that's what's real."

This exchange came back to me, hauntingly, when I toured Stef's collection of Palmah photos, enlarged and framed, decorating— if that's the word for those images of unwashed youths and girls in poses of nervy weariness—the hexagonal foyer of his office complex, in his billionaire's industrial park atop a mountain in the north.

I can hardly say more about Stef than has already been reported. *USA Today* says that his company occupies a prominent position in the world market. The company makes high-precision cutting tools, originally designed by Stef in a wooden shed with a bicycle leaning against it, which image figures in the

PR. His industrial complex houses high-tech factories, a residential village, three museums, a preschool, and a college for entrepreneurs. The college recruits faculty from American business schools, and the students are young people just out of the army. Also, young Palestinian Arabs. Stef gives them the support and facilities that they need to start businesses and run them for five years. In utero, so to speak: he nurtures these start-ups during their critical period. The subsequent birthed companies, some forty-five as of my visit, have, says Stef, ". . . generated around one hundred million. We have created a middle class."

I like his own words best. Stef in *Penthouse:*

> *Our dear American friends: If you really want to help us—don't help us. If you really want to support us—don't do it with money. In all matters relating to Israel's defense, your assistance always has been and will remain a real and reliable hand of friendship; but your gift of money distorts and corrupts us, making us eternally dependent. . . . It's only when you stop supporting us that we will have no choice but to stand on our own two feet. Then, and only then, when Israel earns a living from the fruit of its labors, from its exports to the international marketplace, will it achieve economic independence—the true dream of Zionism.*

"Jerusalem," Stef said, when I told him where I'd been all summer. "Jerusalem is terrible."

His face was dominated by thin, jutting lips. His brows were tufts, and the mustard glow of his eyes belonged to the lore of dragons. I had never seen such in a human mien. He wore—well, with those backlit, single-minded eyes, his clothes weren't noticeable. We talked in a small conference room on the second floor.

"We made a bad mistake taking Jerusalem for the capital. It's like Sarajevo, too much history, old stories; people fall in love

with history because they're tired. Religion's the same. We should
have chosen an empty place."

I thought of Gur's bare hill.

"Ideals must be changed from time to time," Stef declared,
pacing his words, well used to dictation. "We've built parks in
two more locations and now a new one for the Turks. We're
going to make a better Mediterranean. No more Sarajevos."

I could detect, glancing up, the fast-forward flicker in his features as he performed other mental work that didn't require him
to stop talking. Around us was a continual muted twitter of
phones.

"Change can be made only in an empty space and in countries
looking for a new identity. The moment everybody has to get to
one place at the same time, they don't have time to quarrel."

I asked Stef about his past and his role in Hemmed. His family had immigrated from Germany in 1937.

"My father got tired of that maniac's screaming speeches."
His laugh was like the snort of a steam cap. In Hemmed he'd
been a machinist, making parts for the Loretta as well as mortars
and bazookas. "I tested weapons in the field of combat, in the
Negev and the Galilee. Sometimes the product got lost. Sometimes two and two had to equal five."

"Sounds tough," I suggested.

"No one would give a twenty-year-old kid the money to be
chief machinist of an army anywhere else but in this land, where
tsuris created opportunities. I'm happy I went through that.
Made me take responsibility for finding solutions. If I'd gone
to a university, I'd have ended up falling in love with problems
instead. What about you? What's your story?"

He grilled me on my background, and asked what the message
of my poetry was.

"I'm not sure, I mean I'm not sure there is a message . . . or maybe, it's about things changing, things staying intact. . . ."

"Make it less historical." Disapproval, of an objective sort, was abruptly engraved on his face. "This stuff about your father, that's just ego trips. I didn't like it when Bentov left. He should have stayed. We needed him!"

Smarting, flattered, I told Stef that my father had needed the wealth and freedom of the United States. "The Israeli conditions didn't offer him any possibilities."

"Look out the window!" It was a good riposte. "Bentov was impatient. He should have waited fifty years."

"But he wasn't like you, Stef, he wasn't a leader, people got on his nerves, he only wanted to be his own boss—"

"When the goal is clear, difficulties become less important. You don't ask for coffee and cake halfway through a marathon."

"But his goal and your goal were different! He wasn't a Zionist. You took Israel to your heart and he did not."

"Impossible." Stef was thrown. "Not possible. He must have, if he was singing the songs of the Palmah, he must have. He was a kibbutznik!"

I didn't know what to say. The magnate recovered with a smile.

"You should stay in his place."

"There is no work for an American poet, writing in English, who isn't famous, in this country."

"Write something else. Write articles about how we need to create an export economy."

Who are you calling "we," I thought. There was no saying it.

In Stef's blue Volvo, we were driven over to one of the museums, where he had his unintentional revenge for any shock I might have inflicted. He led me to a vitrine on a pedestal, in which, he explained, lay Israel's first rocket. He had built it himself.

"I thought my father built Israel's first rocket, in '48, in the Gray House?"

"No, I did, in '46. This is it."

Inside the vitrine was a steel object the size of a quart of milk. Beside it lay papers, and an ID with a little toothy photo.

"Oh, is that you?"

"That's the age I was when I knew your father. And there is Shai's diagram that he drew in '45, when we began. You see the date. He designed the plans, and I was recommended to him—I had the reputation of a top mechanic, 'golden hands.' This is a model. The original rocket was fired."

"I see. Of course it would have been."

Hefting the handbag, and the nylon briefcase, I thanked Stef for his hospitality and time.

"I'd like to ask you a favor," he said, shaking my hand. He was at least a head taller than I was, and stooped slightly, looking down with his hard gold eyes. "If you find any old photographs of Itzik that you can spare, please send them to me."

I pocketed his card and promised. Walking toward the exit, I turned to wave. Stef still stood by the vitrine, and his rocket.

Returning to Jerusalem, I phoned Shai.

The kitchen felt chilly. The leaves in the balcony door were dark, and the sky between them, a spiritless slate. The pall did not lift when the stove's gas flame whuffed blue under a pot of take-out kubbeh soup from the Kurdish restaurant. None of my talismans were working.

"Shalom," I said into the cold receiver. "I hope I'm not

disturbing you. It's Sharona. I was just visiting Stef, who mentioned you."

"Aha."

"He has a rocket about twenty inches long, which he claims you and he worked on in '46."

There was a rough intake of breath. "That does not seem . . . correct. Too early."

"You designed it, according to him."

During the thoughtful pause on the other end of the line, my soup sent up a simmering aroma; the kubbeh agitated in the pot with comforting thumps. Then Shai passed sentence, in a droll growl that I remember yet.

And whenever I remember Shai's words, his lucid soul stands out from the shades of the recent dead.

"You see, Sharona . . . what comes of research? All the bastards return to us . . . from our forgotten nights of love."

Part Five: Touching the Moon

Twenty

THIS DOOR LEADS onto the moon, as surely as if I'd touched it with my finger and could show you the dust. It's no metaphor: I did touch the lunar surface. This is the last door into memory, the door to the moon.

Steep dark stairs, whose underworld smell of mildew and cooked plastic clings to the roof of my open mouth. I grip the handrail, lower my right foot, and a while passes until my shoe hits the step; then follows the easier task of bringing down my left foot. Each stair is a new venture with its risks of a wobbling knee or a hand's slip. The handrail plunges downward without supports: under my reaching arm is a drop through which I could fall. Slowly, a wan light comes from the taboo side of the basement, Sonia's side, where I am forbidden. I do not look down. Now the black overhang of the upper floor is lifting, as though I went underneath the porch of the world, and another world, my father's, is rising. A wall of brown apothecary jars and large clear bottles, labels affixed in rows, looms higher and higher. I watch the glassware glimmer as I step again and again into the shadowy drop.

. . .

There's Daddy. His woven wire-back chair looks pinched against the generous wave of his shoulders. The castors chirr as his legs, under the table, push him gently back and forth. One hand reaches up, on its own, to adjust the black cone of the crane lamp. His hair ruffs above his collar. A blue reflection, like a window, shines on the bald dome of his lowered head.

I creep to his side. To his table that was once a door. It is covered in plastic scarred by X-Acto knives. My father is writing in a spiral notebook; his wrist rests on the paper. He draws a tall gear, like a Stetson hat. He draws an arrow, then writes: THE WHOLE CROWN COULD BE EXTRUDED FROM STEEL OR ALUMINUM.

Can't he feel the scrutiny I am devoting to his face, his upswept brow with its hooked end? His wrinkled eyelid. Sometimes his cheek is sucked and gnawed, or the corner of his thin lip pulses. Usually, he is singing; his tuneful bass rumbles through half a dozen songs, flowing from between his lips. At the bottom of his range he drones his favorite, the first Hebrew I learn:

Smoke rises from the bonfires and the drums are still.
On the ridge a moon is lighted with a face of blood.

"Daddy?"
He knew I was here.
"Little Pookli. Wait quietly and I show you the ruby laser."

Obediently I wait, then wriggle into my parka, tugging the hood over my head and through its furred arch looking up as he switches off the fluorescent lights. One by one, areas illumined by the tubes blink, and on tables and workbenches, the lathe, belt sander, drill press sink into shadow, into sharing a secret. Yes, a secret's in the air—but what is it, what? There are the sleepy smells of sawdust, mildew.

"Where's the ruby? Can I touch it?"

I'll beg to take it home. He'll forget, and I'll keep it forever.

"No, the ruby is part of the mechanism."

Bearish in his suede jacket and gloves, he unchains the back door and we go into the cold. I skip to keep up with my ruby. Under his arm is a steel box, like the boxed wine my mother brings to a party. But steel, not pretty silver cardboard. The icy gravel crunches. Instead of going to his car, to drive me home, he walks around to the outside wall of his house. A hedge bristles past where we stand, toward the sidewalk; a streetlamp casts a stony light. In our spot, it's dark. My father takes my arms and wedges the laser into them, seating it against my chest. I hold on tight.

"Now press the button," he orders.

Nothing happens. Daddy is laughing. A slat of the siding stares at me with one red eye. I shift, and the eye scoots to another slat. Turning around I watch it vanish, to the sound of his laughter, reappearing a few feet away, bright and baleful, on the garage door. Then on the downspout: a red dot. I'm aiming something that unlike a flashlight beam doesn't shine. Illumination is not its job. The red dot skims the garage roof, back and forth; it should scorch the shingles with a hissing trail, but it makes no sound and does no harm. Over the roof ridge, in a wash of winter stars, the moon floats white and full. The same instant that I dare to aim the laser at the moon, I am swept by icy pinpricks.

Nothing happens.

"Why isn't it there?"

"No, Pookli, the beam is there, only you cannot detect it at such a distance. It is making a dot about half a mile in diameter. On the moon, possibly, you are detecting it, but only with very sensitive instruments."

In the dead cold air, encircled by his arm, aiming the ruby laser at the moon, I imagine my dot. Crossing space, landing so diffuse and faint it makes one pink flash off a dust-grain's facet, on the skin of a dust ocean, inside a giant crater. My red dot is there. My red dot is not-there. It is at once a monumental sign, and a perfect secret. . . .

What was the sign, the secret? What made them one and the same? What was it, which, standing together, we beamed forth into the universal night, in steadfast waves of ruby light?

It was big.

It was ours.

It was detectable only with sensitive instruments.

Twenty-one

I USED TO ASK my father all sorts of questions, practical advice on health, and that's what I'm doing now—because he's not around anymore, I have to use the second best thing, the Idea Books. For a week, since my visit to the doctor, I've been wondering if there's an answer in this notebook. I stroke the brown cover, its glossiness dulled around the corners where the cardboard is like felt. Open it. Here goes . . . you're on, girl. Now let's see if I'm smart enough to puzzle out an answer from among all this intimidating engineering data.

DILATOR—CERVICAL 4/7/70

THE IDEA IS TO MAKE A CERVICAL DILATOR WHICH WILL OPERATE IN A MANNER MORE AKIN TO THE NATURAL FUNCTION OF THE CERVIX + UTERUS.

"The dilator is putting you through college," he used to say, satisfaction mingling with the long-suffering tone in which he talked about my upkeep. The memory is twenty-six years old. Today is the first time I'm paying attention. Yes, I'm raptly attentive, because of what happened to me last week, around the time of the first autumn storm.

. . .

Lights were dimmed to orange circles in the ceiling. On my bare belly, the ultrasound probe slid here and there, warm and greased, with the touch of a cosmetician's fingers. I liked the ultrasound because it's one of those rare times when you're doing the best you can, and the most you ought, by just resting. Over my shoulder, I saw the monitor screen light the technician's face with a somber glow like a three-quarter moon. With the hand that wasn't sliding the probe around, she clicked a mouse pad, causing a few white crosses on the screen to freeze in place, expand, rotate, and freeze again. Her shoulders sloped steeply, her neck was correspondingly elongated, and her head, with its terrier hairdo and granny glasses, seemed outsize compared to her elastic limbs. I liked the tech. After the talk I'd just had with my doctor, she was a distraction.

A gray funnel with an irregular bulge in its wall, that was what the monitor screen made of it. The image reminded me of our quarry in August, when the water level drops down low, exposing the curved shelves of limestone, gray and chalky.

"Looks bumpy in there," I found myself saying. It looked the way it felt. The warm probe slipped toward my left hip, and on the screen the white crosses zoomed like dragonflies.

"Gets your attention," the tech agreed. "There's your big fella, I'd say, oh, about tennis ball–sized. The rest aren't exciting." I liked her sense of humor too. She stuck four crosses compasswise around the tumor and froze them for its portrait.

"The doctor said a baby isn't the greatest idea," I angled. The tech's back, twisted to bridge between the supine patient and the monitor, looked tense, and its owner went on clicking.

"Your ovaries look normal. That's good. You can have a cruddy uterus, but ya gotta have nice ovaries," she said jocularly, sliding down from her perch. She was quite short. "You're done. You can get dressed." On her way out, she turned up the lights in the closet-sized room.

I swung my legs over the examination table's edge and used a napkin to wipe lubricant from my flesh. Barefoot, I tiptoed around to retrieve my jeans, chamois shirt, and the shearling-lined boots that I wore without socks. Until now, I had felt safe inside these clothes. The tech returned, holding a clipboard to her flat chest. She wore a lab coat printed with pink and blue clouds on which yellow babies rolled, kicking up their feet; it made me wistful. She was dressed for the usual ultrasound.

"The doctor says there's nothing I can take," I mentioned, buttoning up my shirt. The tech gave me a fierce smile that cocked her chin at an angle to her long neck, in its pink-cloud-festooned collar.

"One out of four American women have fibroids and, no, we can't give you anything, because there isn't much research on it because fibroids don't kill you. They cause bleeding, and they cause hemorrhage, but they don't direckly kill you. Congress won't hardly fund research for breast cancer, which does kill you. There is a IUD which will halt the growth, not shrink it, just keep it where it is, but your insurance plan won't pay for it because a IUD is a contraceptive and they don't pay for those. Now if you had an enlarged prostate there's a drug we could give you, and that same drug might reduce fibroid tissue, but your plan won't pay for it if it's not for use on a prostate. And don't ask, because you can't afford it."

"Sheesh," I said, looming over my information source and zipping up my parka. She snapped me another smile.

"Write your congressman. I guess I'll be seeing you again

soon," she predicted with a whoofing undertone. My uterus had discomposed her.

After settling accounts with the clinic, I walked out to my car, in the chilly weather. I sat down in the driver's seat and folded my hands, holding the ignition keys in my lap. Thoughts trailed by like animated captions and dissolved. The car shook in a gust. The steering wheel column and dashboard were streaked black where my fingers had brushed the dust. The upholstery was shredding away from the seat's plastic piping. I didn't take care of this car's looks. But I'd relied on its staunch little engine.

I looked into the rearview mirror. Highlights like shining slits shone off my contact lenses, giving a feline look to my eyes, underscored with two lilac rings. Now I hardly registered the usual sigh at those age rings, afloat in the bemused, almost humorous state with which I meet a shock. I ran my mind over the mesmerizing shape of it. I was less myself—that shy, faceless, yet familiar being—by one power that had been truly mine: my compromised fertility.

I'm applying my whole intelligence to my father's design for a cervical dilator. I read his introduction once more.

THE IDEA IS TO MAKE A CERVICAL DILATOR
WHICH WILL OPERATE IN A MANNER MORE AKIN
TO THE NATURAL FUNCTION OF THE CERVIX +
UTERUS. I.E., THE DILATING STIMULUS HAS TO
COME FROM THE INSIDE OF THE UTERUS IN
THE DIRECTION OF THE CERVIX.

The idea is dated 1970, when women were still routinely strapped down in a horizontal position for labor; when few people, except for some ridiculed feminists, took the natural function of women's bodies as a guide to anything. Yet here is my father, in 1970, doing exactly that. He asks: How does the uterus naturally expand its millimeters-wide opening? By sending out a signal to the cervix. When a baby's head needs ejecting, the uterus sends the signal "Dilate," and the cervix obeys.

Then why not make a dilator that causes the uterus to send that signal? A balloon would do it. The doctor slips a collapsed balloon, no more than a rubber thread, into the womb. The balloon slowly fills with water, expanding; and when it is pulled gently back, the uterus sends its baby-coming-through signal to the cervix.

THE ADVANTAGE OF THIS SYSTEM IS THAT
THE UTERUS IS BEING PULLED FORWARD BY
THE DILATOR, RATHER THAN PUSHED BACK IN,
AS WITH THE OTHER METHODS.

He describes the other methods, which push the cervix backward and pry it open from outside, inflicting trauma and cramp—yep—it's just like ramming your head on a sharp corner, only the corner is in your belly and you can't stop hitting it. Of course, the cervix is not permanently damaged; it is merely, as the doctors say, "insulted." But my father's method deliberately follows the natural steps to avoid pain. I remember, he would scold me if I handled a can opener or a pencil sharpener with rough impatience. He would take it, and gently get it working.

"Never force a mechanism," he would say. "Find out what it needs." If I am reading these yellowed notes aright, my father took the same approach toward a woman's body.

Was this who he was all along, and I never gave him credit, never understood his gentle touch, till now? All my life, I've thought of him as a genius. Or as a mad genius. As the elusive father for whom I yearned. I never—not once!—thought of him as compassionate toward women. But the inventor of this device would have had to be. The dilator doesn't save lives. Its purpose is to prevent pain. Coming from a man, it is—dare I think it?—a *chivalrous* invention. Gosh, I just don't believe this . . . it's taken all my life, the twenty-two years since he's been gone, and the twenty-two years I knew him alive, or thought I knew him, to show me what I missed.

Once, I remember, we were talking on the phone. My father was telling me how Mother Nature had visited him in a dream, and shown him the answer to a mechanical problem. It was in my early teens, I think, when this sort of thing was still benign. I asked what she looked like.

"A beautiful lady," he said warmly, "who sits in a golden egg." His voice glowed with love. The love within him.

No cure for fibroids in these notes. I switch off my desk lamp. Stacked before me in the dusk, the Idea Books are like things in a dream, soft-cornered, dim. I think they are still deep in my father's dream of conversing with Mother Nature. Smiling over the page, it is I who feel real—awake, less numbed by my experience at the doctor's, which seems strangely to enliven me, to spring into a power of comprehension. It is easy to forgive this inventor, who sought to spare women insult and pain, for not quite pulling off the role of father. And it is a fine thing to be delivered into forgiveness. I want to call him on the phone and tell him how proud I am . . . I'll hear him say my foolish nickname.

"Little Pookli!" How pleased he sounds, on the other end of the line. He always sounded pleased to hear from me. Whatever else fatherhood meant to him, I was—a nice spot in his day. How could I have overlooked what is such comfort to me now? It never attracted my notice; I'd heard without hearing, all those years. . . .

Then I'll ask about the fibroids that threaten our family line. He's taken aback—it means he's getting old.

"*Yah!* Already you are having such things?! . . . *Nu, nu.* Take a pen. Now listen carefully, because I tell you precisely what to do."

The well-remembered voice stops; if there was an answer, it died with him. Instead of bereft, I feel blessed. I was lucky, oh thank God I had the chance. I was lucky to have opened this notebook. What if I'd gone the rest of my life with the old, automatic disbelief plugging my mind, so I never remembered that all along he'd been happy to hear my voice, my questions? I was lucky, it was by chance, yet everything is changed; and it's funny, how simple and quick such a change can be, just like my ears unpopping after an airplane flight. I can hear, in memory, each note of his natural love, as genuine as any pearl necklace. The love in my father's voice is mine, mine to keep. And always was.

My hair's down and I'm wearing only my fleece robe, but I go through the darkened house into the garage, step into moccasins, and walk outside. The night is mild. The moon is too bright to look at, like a searchlight in a revolving drum. Tree trunks are striped on their moon-facing side, and the branches seem immobilized in cobweb-colored dust. On the asphalt, moonlight layers a deep mottling through which I step gingerly. Dark drifts of dead leaves waft their incense on the night. Coming around the bend where the witchy-limbed blue ash trees overshadow the drive, I hum a little and make up rhymes to keep off the spooks, the way I used to do when I was small. Coconut milk, thick as silk, whose woods these are I think I know, the hunter's moon and then the snow. . . . There, I'm past the bend, and now the drive shines ahead, smooth and calm. It's strange to think that my long female partnership with the moon is coming to an end. Soon, sooner than not, I'll be old. But there will still be nights like this, when it's lucky to be alive. I pass the woodpile beside the barn. Not fifteen feet away, Tom and the dogs materialize in the woods, coming toward me over an intermittent floor of shiny patches. We greet each other with exclamations over the beautiful night. Mix trots up and sits beside me on his poor fixed rump, panting, happy, lower fangs gleaming like the rims of his eyeballs. I can make out the raw silk tint of his coat. Tom holds onto Daisy's leash while she roots in the darkness; he squeezes my hand.

"Ssh—hear him?" he whispers. "Back that way." In the depths of the woods, a great horned owl hoots five times. While we listen for him to call again, the night becomes noisy: a train whistling as it rolls through the junction; the corgis on a neighboring farm, barking off in the distance.

Whoo, hoo-hoo, whuh-hoo-hoo, woofs the owl a second time. The sky looks not black at all, but a milky gray, bottomless.

EPILOGUE

From the Sweet Oak Diary

MARCH 30, 2002
2:00 P.M.

TODAY THE TEMPERATURE dipped back into the twenties and a strange thing happened as well. Tom, who doesn't usually call when my study door is closed, yelled for me to come quick. I pelted into the living room, where he stood to one side of the bay window, peering out. He grabbed my arm and put a finger to his lips. I saw that a thin snow had materialized in the air outside, which had gone gray. On the rocks behind the koi pool, just far enough from the cattails that it stood out, was our visitor.

"That thing's what? Four feet tall?" I hissed.

"At least," Tom whispered. "More. I bet it just ate our koi. I sure hope it didn't just finish eating our koi. Look at him looking at us."

Through the snowy mist, the creature's breast looked like a shield of blue steel. Its head, turned sideways, was a flash of black eye stripe going one way, and a beak to rival Tom's metal cutters going the other way. Then the whole stabbing ensemble of its head, propped on a bend of its neck, moved from right to left, the better to view us. I tried to sneak away for the binoculars, but

Tom clutched my elbow. Too late, I watched the heron's depar-
ture procedures: cranking its wings out, beating them with heavy
strokes that shook the cattails, rising without hurry, and cutting a
figure out of all local proportion in the treetops as it left.

"I bet it ate the koi," Tom muttered oracularly.

I ran out barefoot. In the clear brown pool, small red koi were
cunningly hugging the base of the lily pots. The white one stood
on its head pretending to be a plant. Black-encrusted, orange
Senator, who trusts he's too big to be swallowed whole, lolled at
his ease. I made an okay sign to the window. Then, sensing an
absence, I looked up into the bare branches, and the faint
rhythm of the snowfall ceased as I watched; as if the great blue
heron had brought the snow on his wings, and magically carried
it away again. The scent of snow hung in the air. . . .

It reminded me of a Russian folktale, at first, then of someone, a Russian, whom I knew . . . and when I remembered him, I
thought, that's it! That's how we come back! That's how. My
commander has come visiting.

AFTERWORD

WHEN I BEGAN writing this book, I had the good luck to be a guest at a Passover seder that was as abundant in companionable discussion as it was in delicacies. After someone had finished reading a portion of the Haggadah, our hostess remarked, "It's all about telling, the most important thing is telling." Over the seven-odd years that followed, I often contemplated that aside as I struggled with a difficult question: What is the relationship between memoir and history?

To find an answer, I also had to ask: Whose memoir? Whose history? Who is the author? Well, the author is someone who picked up the thread of her father's life, untimely cut, and wound it into a ball. And, as Blake says, it led her to Jerusalem's wall. Less poetically, I went on a personal search, in the course of which I was granted some inside information about, and possibly insight into, Israel. This does not makes me a historian or a journalist; certainly not a Middle East pundit, Israel maven, or any kind of political or cultural expert. This makes me a writer renting a historian's apartment.

And after many evenings spent over history books, I've come to think that the relationship between memoir and history may be defined, loosely, with a metaphor. Memoir is the dermis of history. It adheres to history, as the dermis adheres to the outer skin. For the dermis and epidermis are not to be separated. Yet the dermis has its own character. It is too sensitive to exposure, just as private memories—the stuff of memoir—cannot always hold up to the cold, dry air of fact, which history is expected to weather. But the dermis shelters the internal organs, as memory enfolds our deepest mysteries.

So my memoir must stick to history, without trying to become it. It must bind history to the interior world: that is, parallel to its duty to history lies its obligation to expression. My memoir must satisfy the reader's curiosity about historical facts without cheating. At the same time, it must be alive with personal feeling, mystery, and dream.

This book's contents, then, range from strict fact where Hemmed's history is concerned to gentle license with my personal tale. For all the events and data related to Hemmed's projects, I have relied primarily on conversations with Hemmed members and the firsthand account *Hemmed Days;* secondarily, on histories. Oral history in the form of conversations imposes certain limits; there is no method for checking, for instance, the words that passed between Yehuda and Weizmann. But where the stories agree (which is almost always the case) and seem probable, I have let them stand.

Hemmed members requested that they appear under pseudonyms. Stef and Shlomo Gur, however, bear their own names—the former by his permission, the latter posthumously—as it would be pointless to disguise them. The curator at Rafael, "Pesah," is the sole instance of a pseudonym standing for a compound character, created from two museum officials; this liberty was also taken out of discretion. Similarly, the book does not identify all the characters with their own faces in the "Passover photograph."

Afterword

This story of Hemmed in '48 is a partial one. I talked chiefly with my father's friends, physicists and engineers from the rocketry group and ballistics laboratory. Hemmed also included groups for chemistry and biology. But these people did not work with my father, and I did not pursue their stories. (The one biologist whom I attempted to draw out made a face and said, "We are the generation that does not talk.") I have also had to be selective with my own material, regretfully leaving out stories such as that of Hemmed's bazooka workshop, manned partly by American volunteers (who shocked the corps by screening blue movies). Or the Hemmed member whom Ben-Gurion set to work reading thousands of letters that poured in from would-be inventors, offering ideas like the giant antiaircraft magnet. And so forth.

With my own life, I have made as free as any memoirist, changing names to protect privacy; sometimes folding memories into one comprehensive scene, or paring down a situation to its gist; and taking to heart the maxim of Michael Ondaatje, in his exquisite memoir *Running in the Family*: ". . . the book is not a history but a portrait. . . ." In the portrait of my father, where I have not used such documents as the Idea Books, I have followed my memories. My intention is not to promote his ideas, which, not being a scientist or a philosopher, I am unqualified to support. I present his ideas because they were what we shared.

In portraying my father's widow, I faced the difficult task of revisiting old, hard feelings; her portrait is necessarily drawn from the painful period that the memoir describes. I can now appreciate other aspects of "Sonia": her successful artistic career, her effective work in keeping my father's ideas before the public, and her devotion to my father in their close partnership. They shared, as my father's letter attests, "really good times." Happily, love endures through the changing perspectives of memory.

A note on Israeli history. In the past three decades, writers ranging from scholars to journalists, of diverse ideological stripes, have rewritten the history of the '48 war, based on archival material that became available in the late 1970s. These writers have discredited national myths about the war and the treatment of Palestinian Arabs. The myth that directly concerns this book is often called "David and Goliath": a defenseless, outnumbered Israel miraculously defeated overwhelmingly superior Arab forces. In fact, historians show that the problem Hemmed struggled with—lack of antitank weapons—was of brief duration, from the outbreak of the invasion on May 15, 1948, till the first truce on June 11. After that, foreign arms deals turned the tide. In hindsight, this was not a lengthy interval. Though during that time, in the first month of the invasion, Israel lost over 1,600 people, 1,200 of them soldiers—fully one quarter of the war's total casualties.

Various histories contributed to the general background of this book, from Walter Laqueur's classic *A History of Zionism* to the works of Yoav Gelber, Avner Cohen, Bernard Avishai, Ilan Pappé, Avi Shlaim, Simha Flapan, among others. What can a memoir add to their facts, figures, and corrections of the record? Hemmed members told me, over and over, "We had nothing." In hindsight, it is true that "nothing" was just a gap in the flow of arms. But in the spring of '48, the Hemmed scientists did not know that they were looking at a gap. What they saw was an abyss. This is the perspective that I have sought to re-create, not as history, but as memoir—as the dermis, the layer that, above all, is glimpsed through wounds.

I was not compelled to write this memoir for political reasons; it makes no prescriptions for the Middle East. The reader may label my own politics as he or she sees fit, but I can't help feeling that such a label is neither here nor there. This is an American story: a quest for roots across the sea, and a return home. It is written out of a very personal desire to remember my father and his friends; their personalities, their world, their war, their ideals, their achievement as they understood it. Because every human generation is, as Misha rebuked, "the generation that forgets," unless we invent memory anew.

NOTES

CHAPTER ONE

4 which is classified: Avner Cohen mentions the classified status of Hemmed (in his version, HEMED) in his article "Before the Beginning: The Early History of Israel's Nuclear Project 1948–1952," *Israel Studies* 3, no. 1 (Spring 1998): 113. Readers who wish to know more about Hemmed's postwar development into Israel's nuclear research establishment would be well advised to read Cohen's groundbreaking study *Israel and the Bomb* (New York: Columbia University Press, 1998).

6 "an extraordinary, brilliant group": Mr. Shimon Peres talked with me briefly during his visit to Toledo, Ohio, in 1997.

6 *Stalking the Wild Pendulum*: Itzhak Bentov, *Stalking the Wild Pendulum: On the Mechanics of Consciousness* (Rochester, Vt.: Destiny Books, 1988; first printing, E. P. Dutton, 1977).

CHAPTER TWO

14 "The disk is a thing called a hologram": My father's ideas, as they appear in this chapter, also appear in his book with some of the same phrasing and images, such as the can of peaches (*Pendulum*, p. 28). There are two reasons for this. First, my father used many of these phrases and images to describe his ideas to me long

before they went into print: by teaching, he learned the most effective words to convey his ideas. But I have also combined some of *Pendulum*'s language with my own memories in order to convey, as vividly, accurately, and succinctly as possible, the gist of discussions that spanned years. In effect, I have sometimes capitalized on his excellent work of summary in *Pendulum* to give the reader a concentrated version of our typical Saturday discussions during the years that he was formulating his central ideas.

CHAPTER THREE

26   an attack that sank the Egyptian navy flagship: The operation is described by Mike Eldar in *Shayetet 13: Sipuro shel ha'commando ha'yami* [Flotilla 13: The Story of Israel's Naval Commandos] (Tel Aviv: Ma'ariv Book Guild, 1993), pp. 142–51. This story may be appreciated best by a visit to Haifa's maritime museum, where the reader can see one of the scary little boats known as "pigs" or "cigars." My knowledge of the role "Plafchik" played comes from conversations with Hemmed members.

CHAPTER FOUR

43   a slim, privately published memoir: Gideon Yekutieli, *Yemei Hemmed* [Chemed Days] (Rehovot: Weizmann Institute of Science, 1996), p. 2. I have drawn on this account for anecdotal details throughout this chapter.

47   "I almost wish that it had been as simple as that": Walter Laqueur, *A History of Zionism* (New York: Schocken Books, 1978), p. 187.

48   All over the country, on walls and kiosks: Images and items in this passage are drawn from Mordecai Naor's chronicle *The Twentieth Century in Eretz Israel: A Pictorial History* (Cologne: Könemann Verlagsgesellschaft mbH, 1998), pp. 216–89.

48   From a community of 600,000, some 27,000 enlist: See Bernard Avishai, *The Tragedy of Zionism* (New York: Farrar, Straus and Giroux, 1985), pp. 155–56. Avishai mentions that some 1,000 Palestinian Jews also served in the Free French Brigade, "of whom only 45 survived the defense of Bir Hacheim" (p. 155).

49 *I cannot understand why this course has been taken:* Laqueur, *History of Zionism,* pp. 510–11.

50 Despite the noose tightening on the neck of European Jewry: Ibid., p. 509.

50 The *Struma,* turned back and torpedoed: Naor, *Twentieth Century,* pp. 217, 226, 236, 245.

50 Britain pursues this policy: See Ilan Pappé, *The Making of the Arab-Israeli Conflict, 1947–1951* (London: I. B. Tauris & Co., Ltd., 1994). "The British attempted to prevent Jewish immigration into Palestine until the last day of the mandate, seeing this as the only way to maintain law and order; continued immigration would have caused, so they thought, extreme and violent reactions not only from the Palestinians but also from the Arab world at large" (p. 21).

50 My historian landlord suggests Realpolitik: "The Arabs were many and the Jews were few. Precisely in view of the coming war, Arab goodwill had to be won" (Laqueur, *History of Zionism,* p. 524).

50 Or responding to Palestinian Arabs: See the citation of Walid Khalidi in Simha Flapan's *The Birth of Israel: Myths and Realities* (New York: Pantheon Books, 1987): "'The Palestinians failed to see why they should be made to pay for the Holocaust. . . . They failed to see why it was *not* fair for the Jews to be in a minority in a unitary Palestinian state, while it *was* fair for almost half of the Palestinian population—the indigenous majority on its own ancestral soil—to be converted overnight into a minority . . .'" (pp. 57–58). This quote explains the response of Palestinian Arabs to the U.N. partition plan. It is helpful as well for understanding the Palestinian Arab objection to Jewish immigration, demographics being key to these related issues.

50 The authorities claim that it is dangerous: Laqueur, *History of Zionism,* p. 536.

51 *We have not failed:* Ibid., p. 531.

51 *the homeland is not like the Diaspora:* Naor, *Twentieth Century,* p. 212.

52 The same period that sees your 1st and 2nd infantries: Ibid.; items included in this and the following passage on pp. 216–65; Black Saturday, p. 246. See also Avishai, *Tragedy of Zionism,* p. 170. Laqueur puts the number of bullets for which Weizmann's bodyguard was sentenced at two (*History of Zionism,* pp. 535–36).

54 "did not want too many of them in New York": Conor Cruise O'Brien, *The Siege: The Saga of Israel and Zionism* (London: Paladin, Grafton Books, 1988), p. 267. Apropos of Bevin's attitude, Pappé discusses Bevin's desire to use the postwar Anglo-American alliance "in order to advance the chances of a solution in Palestine . . . he believed that such an alliance would mitigate the influence of American Jewry on the White House, which he saw as the primary cause for the lack of unity in the two powers' policy towards Palestine" (*Making of the Arab-Israeli Conflict*, p. 11). Pappé points out that most historians would regard Bevin's notions of Zionist influence as exaggerated.

54 *If you think of bringing the redemption nearer:* Laqueur, *History of Zionism,* p. 575.

55 When Misha insisted that not one Arab: See the Afterword.

55 Privately reserving the thought that in the factionalized Arab community: The fragmentary nature of Palestinian Arab hostilities, with some localities active and some quiet, was recognized and played a role in Jewish discussions of retaliatory policy, according to Yoav Gelber, *Palestine 1948: War, Escape and the Emergence of the Palestinian Refugee Problem* (Brighton: Sussex Academic Press, 2001), p. 63.

55 Recent historians convincingly show: See the Afterword.

55 War #1 and war #2 merge: See Gelber, *Palestine 1948:* ". . . by mid-January 1948 the main Palestinian effort was clearly directed against Jewish transportation" (p. 26). Before the first truce on June 11, "Ben Gurion . . . received letters and telegrams from Jerusalem painting a disturbing picture of a desperate community" (p. 148).

For a sensitive firsthand account of the battle waged on Palestine's roads, and around Jerusalem in particular, I recommend the prizewinning diary of a Palmah fighter, Hadassah Avigdori-Avidov, *Baderech shehalachnu: meyomana shel melavat shayarot* [Convoy Escort] (Tel Aviv: Israel Ministry of Defense, 1988; 2nd ed., 1989).

56 It was loved for its intrinsic beauty: See, for example, Larry Collins and Dominique Lapierre, *O Jerusalem!* (New York: Pocket Books/ Simon & Schuster, 1972), pp. 554–56. This account cred-

its "one of the world's most distinguished physicists, Joel Racah," with building the device, but does not mention the existence of Hemmed, nor its subsequent development of the Hollow Charge.

57 She notes the dignified Sudanese servants: Malkah Raymist, *The Stiff-Necked City* (Jerusalem: Gefen Publishing House Ltd., 1989), p. 119. In the following sections, I have used Ms. Raymist's lively details and anecdotal descriptions, with their markedly colonial color and flavor, to incorporate her into this tale as a character. I have not used this source for otherwise uncorroborated historical facts or for their interpretation.

57 "Round little wounds": Ibid., pp. 199–200.

58 At 6:30 a.m. on a February morning: Ibid., pp. 59–65.

58 Some tall blond Englishmen got out: According to Gelber, "The perpetrators were British defectors in Abd al-Qadir al-Hussayni's service" (*Palestine 1948*, p. 24).

60 Britain stays officially ignorant of Israel's existence: A front-page article in *The New York Times* on May 19, 1948, summarized the British position and tied Britain's nonrecognition of Israel to its continuing military and financial aid to the Arab Legion: "Britain will not withdraw British officers, the monetary subsidy or the supply of arms to Trans-Jordan's Arab Legion unless the United Nations decides that the Arabs are acting illegally, a Foreign Office spokesman said tonight. . . . So far as the state of Israel is concerned, Britain officially does not know that it exists." Herbert L. Matthews, "British Firm on Legion Aid Till U.N. Rules on Arab Acts," datelined London, May 18, 1948.

60 Egypt's largest newspaper avers that the problem of recognizing Israel: A report in *The New York Times,* May 17, 1948, p. 3, cited *Al Ahram* on the view of Egyptian officialdom. "Al Ahram, influential Arabic language newspaper, said that . . . official circles were astonished [by President Truman's recognition of Israel] and added: 'Never before in history did such a recognition come so quickly. It will go as quickly—when existence of the new State ends.'" In Associated Press, "U.S. Action on Israel Criticized in Egypt; Recognition Is Said to Ignore Arab Nations," datelined Cairo, May 16, 1948.

60 Israel's supply of rifles, machine guns, and light and medium mortars: This summary of arms is drawn not from my conversation with Shlomo Gur, but from Gelber, *Palestine 1948*, p. 13. This is the information behind Gur's emotional expression, to me, of the situation as he saw it—see Chapter Thirteen. What Gur feared, however, was not quite what threatened. Sources used in this book agree that the Arab nations' reasons for invasion were politically complex, and that neither their goals nor their means matched their rhetoric. In Avi Shlaim's words, "the combined and simultaneous Arab invasion turned out to be less well-coordinated, less determined, and less effective than Israel's leaders had feared" (in *The War for Palestine: Rewriting the History of 1948,* ed. Eugene L. Rogan and Avi Shlaim [Cambridge: Cambridge University Press, 2001], p. 93). Gelber describes the clash of Arab rhetoric with Jewish perception: "Despite the wild rhetoric that had preceded and accompanied the invasion, the invaders' goal was not and could not be 'pushing the Jews to the Mediterranean.' The purpose of this propagandist slogan was mobilizing domestic support. . . . The *Yishuv*'s [Palestinian Jewish community's] comprehension of the Arab onslaught was totally different. Against the backdrop of the Palestinians' violent opposition to the Zionist enterprise . . . and the Arab states support for their struggle . . . the *Yishuv* perceived the peril of an Arab invasion as threatening its very existence" (Gelber, *Palestine 1948*, p. 137).

61 Of all the visions he has nursed into realities: Dan Kurzman, *Ben-Gurion, Prophet of Fire* (New York: Simon & Schuster/Touchstone Editions, 1984), p. 303.

62 ". . . the entire scientific basis of the Institute": My translation. As cited in an article by Dr. Ephraim Katzir, "Reshito shel Hamehkar Habithoni—Ben-Gurion Vehahemed," in "David Ben-Gurion Vehitpathut Hamada Be Ysra'el—Yom Iyun Bemeleat Meah Shanah Leholedet David Ben-Gurion," from a symposium held on April 23, 1987. Jeruselem: Israel National Academy of Sciences, 1989, pp. 25–42; citation on p. 36 of Katzir's article.

CHAPTER FIVE

65 On July 6, 1995, I called Kibbutz Shoval: For a finely written insider's account of Kibbutz Shoval's founding and life throughout the '48 war, I recommend *Hai Ben Hahayim,* published under the name of Kibbutz Shoval by Sifriyat Poelim, Tel Aviv, 1959.

CHAPTER SIX

82 "FIRST INSTRUMENT THAT'S FULLY CONTROL-LABLE": John Kolb, "Product Profile: Elastic linkage is heart of steerable catheter," *Product Engineering,* September 8, 1969, pp. 19–21.

CHAPTER SEVEN

89 a videotaped TV show from 1977: The depiction of the TV show in this chapter combines memories of my father's first appearance on the program *New Heaven, New Earth* with a later appearance, which the reader can also view on the video *From Atom to Cosmos: Evolution of Consciousness as a New Model of the Universe,* copyright Mirtala, 1990, produced by Video Information Services, 2321 Abbot Kinney Blvd., Suite 201, Venice, CA 90291. This video presents Itzhak Bentov's last TV appearance together with lectures by Mirtala for a user-friendly guide to Bentov's ideas.

91 His book was called *Stalking the Wild Pendulum:* See notes to Chapter One. In this chapter, I represent only part of the ideas in *Pendulum,* the ones that interest me the most. As I've said elsewhere, the reader should note that my presentation of my father's ideas does not constitute an endorsement of them; I am not qualified to speak about their legitimacy except insofar as I can give the general impressions of an educated layman.

93 According to my favorite biology book: Lynn Margulis and Dorion Sagan, *What Is Life?* (Berkeley: University of California Press, 1995), p. 8.

95 Yet it is hard to imagine more complex: Ibid., pp. 101–30.

Notes

95 Charles Darwin once noted: Louis Menand, *The Metaphysical Club: A Story of Ideas in America* (New York: Farrar, Straus and Giroux, 2001), p. 210.

CHAPTER EIGHT

103 "We don't want to be 'ourselves'": This scene and the subsequent scene of the "play" in this chapter consist of my loose improvisation on the three-act tragedy *Be'arvot Hanegev* by Yigal Mossinzon (Tel Aviv: N. Twersky Ltd., 1949), which was produced by HaBimah, the famous theater company whose name I have translated as the Dais. I have attempted to convey the sense of '48 in a way that would avoid the quaintness of Mossinzon's play, by reflecting the way that many Israelis, as I know them, still feel and think—and, of course, the way that I feel and think about Israel; in other words, by reworking the '48 material not in terms of realism but in terms of self-conscious reimagination. What interests me about Mossinzon's play is not that it typified the war—it didn't—but rather that it embodied the way people felt and thought about the war. In the café scene, the encounter between Alterman and Rovina is imaginary, but Shlomo Gur did spend time at Café Cassit with Alterman (see notes to Chapter Thirteen), and Rovina did act in *Be'arvot Hanegev*.

104 *When the Brits left, they imposed an arms embargo:* The embargo was imposed by the U.N. Security Council in May 1948, and was initiated jointly by Britain and the United States. It also banned the supply of arms to the Arab combatants (Pappé, *Making of the Arab-Israeli Conflict*, p. 142; Gelber, *Palestine 1948*, p. 14).

106 *Until the foreign arms deals came through?:* See the Afterword. Sources used in this book agree that the armaments of the invading Arab forces were superior to Israel's until clandestine arms purchases tipped the balance of firepower, after the first truce on June 11, 1948. See, for example, Gelber, *Palestine 1948*, p. 12; Pappé, *Making of the Arab-Israeli Conflict*, p. 52; Flapan, *Birth of Israel*, p. 198.

106 *The Egyptians came north:* It seems the Egyptians did come as close as thirty miles, but they were not headed for Tel Aviv: "The Israeli

high command mistakenly assessed that the Egyptians were head-
ing for Tel Aviv, but the invading force turned east to make contact with their troops in Hebron" (Gelber, *Palestine 1948*, p. 141).

108 "The main business of the group of physicists": Yekutieli, *Yemei
Hemmed*, p. 14. I have drawn on this account throughout the
chapter.

111 *It is not the tank that will win:* Netanel Lorch, *The Edge of the Sword: Israel's
War of Independence, 1947–1949* (New York: Putnam, 1961), p. 207.

115 Six hundred shells per hour: Ibid., pp. 216–17.

CHAPTER TEN

131 The ancient Jewish Quarter sealed inside the Old City walls: The
material on the siege of the Jewish Quarter in this chapter is
drawn from three sources: first, "Aia"'s recollections; second,
Collins and Lapierre, *O Jerusalem!*; third, the chapter by Netanel
Lorch, entitled "Ha'rova Ha'yehudi Be'matzor U'bakrav," in
Ha'rova Ha'yehudi Ba'ir Ha'atika Yerushalayim [The Jewish Quarter in
Jerusalem], ed. Mordecai Naor (Jerusalem: Ha'hevra Le'shikum
U'lepituah Ha'rova Ha'yehudi Ba'ir Ha'atika Be'Yerushalayim,
Ltd., 1987), pp. 244–62.

CHAPTER ELEVEN

156 Who put the tribbles in the quadrotriticale?: This is a line from
the *Star Trek* television series created by Gene Roddenberry and
produced by Paramount Pictures. The line occurs in the second-
season episode entitled "The Trouble with Tribbles," written by
David Gerrold, directed by Joseph Pevney.

CHAPTER THIRTEEN

163 the English initials H.Z.M.N.: From an untitled manuscript by
Dr. Amos Carmel, p. 63. I am much indebted to Dr. Carmel for
sharing his time and work. The initials stand for "Harbei Zarot,
Me'at Nahat," i.e., "Much trouble, little satisfaction."

163 he was a one-man Israel: For a portrait of Shlomo Gur, including photographs, see the article by Yosef Argaman, "V'et
Ha'ikar, Haverim, Od Lo Amarnu," in *Ba'mahane,* May 31, 1990, pp.
23–27.

CHAPTER FIFTEEN

196 We pored together over a book about Rafael: See Munya M.
Mardor, *RAFAEL: Be'netivei Ha'mehkar Ve'ha'pituah Le'bithon Yisrael*
(Israel Ministry of Defense, 5th ed., 1988), pp. 205–95.

202 "When inspiration came": James Gleick, *Chaos: Making a New Science*
(New York: Penguin Books, 1988), p. 175.

204 "To be human is to exist in language": Fritjof Capra, *The Web of Life*
(New York: Doubleday/Anchor Books, 1996), p. 290.

204 "[T]he brain does not own any direct copies": Gleick, *Chaos,*
pp. 163–64.

CHAPTER SIXTEEN

209 I recalled what Shai had written: See Yekutieli, *Yemei Hemmed,*
pp. 27–30.

Note on the Loretta: The curator of the Rafael Museum has informed me, since the writing of this book, that the museum staff
mistakenly identified the cannon in the photograph as my father's
work. It has a two-inch-bore diameter, while the cannon that my
father designed was a one-inch-bore diameter. According to the
museum, my father's Loretta was meant to be fired from the
shoulder, which, they tell me, was a revolutionary idea. They have
not found the prototype and are currently searching for information in order to reconstruct it.

CHAPTER NINETEEN

234 *USA Today* says: "Back to the Future: Military-based industries
decline," *USA Today,* June 6, 1995; in "Our World" supplement,
p. iii, produced by United World Ltd., Inc.

235 ". . . generated around one hundred million": Ibid.

235 *Our dear American friends*: Michele Chabin, "Stef Wertheimer: Eye on the Future," *Lifestyles Magazine,* Summer 1994, p. 50, citing *Penthouse* of March 1990.

AFTERWORD

260 Israel lost over 1,600 people: Gelber, *Palestine 1948,* p. 148.

260 What they saw was an abyss: While challenging the David-and-Goliath myth, historians agree on the hardships of the first round of the invasion, before the truce on June 11. Their words help us to feel the atmosphere in which Hemmed worked, while presenting the historical situation without the veil of the myth. "'The first days of the state's existence were filled with horror,'" Flapan writes (citing Michael Bar-Zohar; Flapan, *Birth of Israel,* p. 198). "Jews who had been celebrating on the night of May 14 were mustered into militia units the following morning. Newly landed immigrants were taken directly from their ships to military bases hastily constructed from corrugated tin and barbed wire. Only after war had raged for several weeks did it gradually become clear that Jewish forces were gaining the upper hand," writes Avishai (*Tragedy of Zionism,* p. 179). Pappé points out that "the men and women in the street did not share the knowledge or the confidence of the leadership, while within the élite itself not everyone was equally sanguine about the outcome of the struggle" (*Making of the Arab-Israeli Conflict,* p. 56). Shlaim states that "during this period, the IDF was locked in a battle on all fronts, against the five invading armies. The IDF had numerical superiority in manpower . . . but it suffered from a chronic weakness in firepower, a weakness that was not rectified until the arrival of illicit arms shipments from the Eastern bloc during the first truce. The sense of isolation and vulnerability was overwhelming" (*War for Palestine,* p. 90). Taking a broader view, Ehud Luz comments on the sense of dire necessity that was generally felt, and perhaps lurked in the background of Hemmed's perceptions, in '48: "British policy in Palestine, the Arab refusal to recognize the

Jewish people's right to the country, and, above all, the destruction of European Jewry gave rise to the feeling that the Jews had no alternative but to fight to the death" (*Wrestling with an Angel: Power, Morality, and Jewish Identity* [New Haven, Conn.: Yale University Press, 2003], p. 87).

ACKNOWLEDGMENTS

I wish to thank the following institutions for their support during the research and writing of *The Book of Telling*: the Ohio Arts Council, the Council for the Humanities at Princeton University, the Institute for the Study of Culture and Society at Bowling Green State University, the Memorial Foundation for Jewish Culture, the James Heekin Foundation, the Toledo Museum of Art, and the Arts Commission of Greater Toledo.

I am deeply grateful to Alan Lightman for having read the memoir in several drafts, and for having provided valuable guidance as well as encouragement. Alice van Straalen, my editor, has nurtured the book with great insight, dedication, and tact. Joseph and Ilana Zur were the most beneficent of consultants. Sincere thanks also go to the following people for advice and support at various stages of the work: Steven Copeland, Robert Fagles, Anthony Grafton, Thomas Klein, Fern Levitt, Askold Melnyczuk, Esther Niv-Krendel, Vivian Patraka, David and Judy Weinberg, and Froma Zeitlin. My mother supplied me with treasures of memory and reflection. My husband helped me get through the long labors of this project by reminding me that I was "growing an oak tree, not a tomato plant."

Above all, I am indebted to the former members of Hemmed, who so generously shared their story and helped me to tell it. You know who you are.

University of Nebraska Press

HOUSES OF STUDY
A Jewish Woman among Books
By Ilana M. Blumberg

To learn was to live, and to learn well was to live well. This was the lesson of both cultures of the Modern Orthodox Jewish world in which Ilana Blumberg was educated, with its commitment to traditional Jewish practice and ideas alongside an appreciation for modern, secular wisdom. But when the paths of Jewish tradition and secular wisdom inevitably diverge, applying this lesson can become extraordinarily tricky, especially for a woman. Blumberg's memoir of negotiating these two worlds is the story of how a Jewish woman's life was shaped by a passion for learning; it is also a rare look into the life of Modern Orthodoxy, the twentieth-century movement of Judaism that tries to reconcile modernity with tradition.

ISBN: 978-0-8032-1367-8 (cloth)

FORTUNE TELLER'S KISS
By Brenda Serotte

There was always the incantation: "Whoever wishes you harm, may harm come to them!" And just in case that didn't work, there were garlic and cloves to repel the Evil Eye—or, better yet, the dried foreskin from a baby boy's circumcision, ground to a fine powder. But whatever precautions Brenda Serotte was subjected to, they were not enough. Shortly before her eighth birthday, in the fall of 1954, she came down with polio—painfully singled out in a world already marked by differences. Her bout with the dreaded disease is at the heart of this poignant and heartbreakingly hilarious memoir of growing up a Sephardic Jew among Ashkenazi neighbors in the Bronx.

ISBN: 978-0-8032-4326-2 (cloth)

HOLOCAUST GIRLS
History, Memory, and Other Obsessions
By S.L. Wisenberg

This bracing and vivid collection of essays gives voice to what some American Jews feel but don't express about their uneasy state of mind. In confrontation with this self-consciousness, S.L. Wisenberg is both engaged and urgent. These essays creatively, and sometimes audaciously, address the question of what it means to be an American Jew trying to negotiate overlapping identities— woman, writer, and urban intellectual in search of a moral way.

ISBN: 978-0-8032-9866-8 (paper)

Order online at www.nebraskapress.unl.edu or call 1-800-755-1105.
Mention the code "BOFOX" to receive a 20% discount.